Stochastic Processes
and Estimation Theory
with Applications

Stochastic Processes and Estimation Theory with Applications

Touraj Assefi

Jet Propulsion Laboratory
California Institute of Technology
Pasadena, California

A WILEY-INTERSCIENCE PUBLICATION

JOHN WILEY & SONS

New York • Chichester • Brisbane • Toronto

Copyright © 1979 by John Wiley & Sons, Inc.

All rights reserved. Published simultaneously in Canada.

Reproduction or translation of any part of this work beyond that permitted by Sections 107 or 108 of the 1976 United States Copyright Act without the permission of the copyright owner is unlawful. Requests for permission or further information should be addressed to the Permissions Department, John Wiley & Sons, Inc.

Library of Congress Cataloging in Publication Data

Assefi, Touraj, 1941-
   Stochastic processes and estimation theory with applications.

   "A Wiley-Interscience publication."
   Includes bibliographical references and index.
   1. Stochastic processes. 2. Estimation theory.
I. Title.
QA274.A84   519.2   79-17872
ISBN 0-471-06454-8

**To the Memory of
Professor Nasser E. Nahi**

PREFACE

This book presents an introductory account of stochastic processes and estimation theory with applications. It is primarily intended for first-year graduate and advanced senior level students and practicing engineers and scientists whose work requires an acquaintance with the theory. The subject matter has evolved from a course given at the graduate level in the Department of Electrical Engineering at the University of Southern California.

The mathematical background assumed of the reader includes concepts of elementary probability theory, the ability to use Fourier and Laplace transforms, and an understanding of the basic ideas of linear system theory. Familiarity with linear algebra is helpful but not essential. There is, in general, no substitute for a rigorous mathematical treatment; however, it is felt that the concepts and the important ideas to be presented may be obscured if too many mathematical details are included. Nevertheless, the book is not a "cookbook"; the definitions and theorems are carefully stated.

The approach to and coverage of the material found here were heavily influenced by the author's practical experience with problems encountered at the Jet Propulsion Laboratory concerning pointing accuracies of science instruments for various spacecraft. It is, therefore, hoped that the book will be useful to a large class of engineers and scientists working in the areas of guidance and control, communications, or other disciplines involving stochastic processes, estimation theory, and image enhancement.

To make the book self-contained, the first chapter reviews the fundamental concepts of probability that are required to support the main topics. The appendices discuss the remaining mathematical background. The reader is advised to review the appropriate sections before attempting the problems at the end of each chapter. There are many examples scattered throughout the text, and the problems at the end of each chapter must be considered an integral part of the material. It is emphasized that the notation is generally independent from one chapter to the next.

I wish to thank George Pace and Walter Havens for their encouragement. I would like to acknowledge John Fowler and Patrick Mak for their excellent comments. Last, but not least, I wish to thank Michael Griffin and George Jaivin for reviewing the entire book and making it more readable.

Touraj Assefi

Pasadena, California
August 1979

Contents

1 REVIEW OF PROBABILITY 1

   1.1 Introduction 1
   1.2 Sample Space, Events, and Basic Concepts of Probability . . 1
   1.3 Conditional Probability, Total Probability, Bayes' Theorem,
        and Statistical Independence 2
   1.4 Random Variables and Probability Distribution and
        Density Functions 4
   1.5 Functions of Random Variables 10
   1.6 Some Useful Definitions and Concepts 16
   1.7 Normal Distributions and Characteristic Equations 25
   1.8 The Characteristic Function 29
   1.9 Definition Extended to Random Vectors 30
        Exercises 33

2 STOCHASTIC PROCESSES 36

   2.1 Introduction 36
   2.2 Definitions and Examples 36
   2.3 First-Order Statistics 40
   2.4 Second and Higher Order Statistics 45
   2.5 Stationary Processes 48
   2.6 Continuity and Differentiability 53
   2.7 Ergodicity and Stochastic Integrals 56
   2.8 Stochastic Integrals in Quadratic Mean 56
   2.9 Definition of Ergodicity 63
   2.10 Special Processes with Independent Increments 64
        Exercises 70

**3 POWER SPECTRUM OF
   STATIONARY PROCESSES** 74

   3.1 Classification of Systems 74
   3.2 Frequency Spectra and Fourier Transforms 80

	3.3	Power Spectra	83
	3.4	Major Result	93
	3.5	Input-Output Relations	95
	3.6	Input-Output of Multiple Terminals	97
	3.7	Sampling Theorem	100
	3.8	Summary of Some Useful Results	107
	3.9	Ideal Low-Pass Signals	108
	3.10	Representation of Band-Pass Processes	110
		Exercises	117

4 ESTIMATION THEORY . . . 120

	4.1	Introduction	120
	4.2	Systems and Modeling	121
	4.3	Mean-Square Estimation	123
	4.4	Linear Estimate	126
	4.5	Orthogonality Principle	128
	4.6	Linear Mean-Square Estimate of Continuous Stochastic Signals	134
	4.7	The Wiener-Kolmogorov Theory	135
	4.8	Optimum Causal Systems	145
	4.9	Matched Filtering	154
	4.10	Kalman-Bucy Filtering	157
	4.11	Combination of Unbiased Estimators	183
	4.12	Discrete Smoothing	184
	4.13	Nonlinear Estimation	185
	4.14	Reformulation of Kalman Filtering	190
	4.15	Discussion and Concluding Remarks	191
		Exercises	193

5 APPLICATION OF ESTIMATION THEORY TO IMAGE RESTORATION . . . 202

	5.1	Introduction	202
	5.2	Spectral Factorization	204
	5.3	Recursive Image Estimation	208
	5.4	Partial Randomization	228
	5.5	Conclusions	243

APPENDICES

	A.	Dirac Delta Function	244
	B.	Vector Spaces and Matrices	249

C.	Fourier and Bilateral Laplace Transforms and Their Inversions	263
D.	A Special Vector Space	271
E.	State Variables	276

REFERENCES 283

INDEX 289

CHAPTER 1
REVIEW OF PROBABILITY

1.1 INTRODUCTION

The concept of probability is used in a wide variety of scientific fields, such as genetics, control, communication, econometrics, and many others. In what follows the fundamental concepts of probability are discussed. References [1]–[10] were utilized in the composition of this chapter.

1.2 SAMPLE SPACE, EVENTS, AND BASIC CONCEPTS OF PROBABILITY

1.2.1 Sample Space

Consider an experiment denoted by $\mathscr{E}$. By sample space, we mean the set of all outcomes of $\mathscr{E}$, which is denoted by S. The set S is also called the universal set.

Example 1

Let $\mathscr{E}$ be the experiment of tossing a die and observing the number shown on top. The sample space S is given by:

$$S = \{1,2,3,4,5,6\}$$

1.2.2 Events

An event A is a subset of S, i.e., A is a set of some outcomes which are members of S. Note that if A and B are events, so are $A \cup B, A \cap B$, etc.

Definition 1

Two events A and B are mutually exclusive if there is no way that they can occur simultaneously, i.e., $A \cap B = \phi$, where ϕ denotes the empty set.

1.2.3 Basic Concepts of Probability

Let S be a sample space associated with the experiment $\mathscr{E}$. With each event A we associate a real number denoted by $P(A)$ and define it as the probability of A. The following conditions must be satisfied:

(1) $0 \leqslant P(A) \leqslant 1$

(2) $P(S) = 1$

(3) If $A \cap B = \phi$, then

$$P(A \cup B) = P(A) + P(B)$$

(4) If $A_1, A_2, \ldots$, are mutually exclusive events, then

$$P\left(\bigcup_{i=1}^{\infty} A_i\right) = P(A_1) + P(A_2) + \ldots + \ldots = \sum_{i=1}^{\infty} P(A_i)$$

1.2.4 Some Important Results

The following conditions are true and are left as exercises:

(1) $P(\phi) = 0$

(2) $P(\overline{A}) = 1 - P(A)$, where $\overline{A}$ is the complement of A

(3) $P(A \cup B) = P(A) + P(B) - P(A \cap B)$

1.3 CONDITIONAL PROBABILITY, TOTAL PROBABILITY, BAYES' THEOREM, AND STATISTICAL INDEPENDENCE

1.3.1 Conditional Probability

Let A and B be two events. Then $P(A|B)$ is denoted as the probability of event A such that B has occurred and is defined as:

$$P(A|B) = \frac{P(A \cap B)}{P(B)}, \text{ if } P(B) \neq 0 \qquad (1.1)$$

1.3.2 Total Probability and Bayes' Theorem

Given a sample space S associated with the experiment $\mathcal{E}$ and given events $A_1, A_2, \ldots, A_k$, we say $A_1, A_2, \ldots, A_k$ represents a partition if the following conditions are satisfied:

(1) $A_i \cap A_j = \phi$, if $i \neq j$

(2) $\bigcup_{i=1}^{k} A_i = S$

(3) $P(A_i) > 0$, for all $i = 1, \ldots, k$

Now, let A and B be events. Then we can easily show that:

$$P(B) = P(B|A_1) P(A_1) + P(B|A_2) P(A_2) + \ldots + P(B|A_k) P(A_k)$$

$$= \sum_{i=1}^{k} P(B|A_i) P(A_i) \qquad (1.2)$$

The above result is called the theorem of total probability.

Utilizing the definition of conditional probability and taking advantage of Eq. (1.2), we now get:

$$P(A_j|B) = \frac{P(A_j \cap B)}{P(B)} = \frac{P(B|A_j) P(A_j)}{\sum_{i=1}^{k} P(B|A_i) P(A_i)} \qquad (1.3)$$

The above result is called Bayes' theorem.

Example 2

An electronic company producing transistor radios has three plants producing 15%, 35%, and 50% of the entire output, respectively. Assume the probabilities that a radio produced by these plants is defective are 0.01, 0.05, and 0.02, respectively. If a radio is chosen at random from the entire company, what is the probability that it is defective?

Solution

Let

$$B = \{x \text{ (radio)}: x \text{ is defective}\}$$
$$A_i = \{x: x \text{ is chosen from plant } i\}$$

Using Eq. (1.2) yields:

$$P(B) = \sum_{i=1}^{3} P(B|A_i) P(A_i) = 0.01 \times 0.15 + 0.05 \times 0.35 + 0.02 \times 0.5 = 0.029$$

Example 3

Assume a radio chosen at random is found to be defective. What is the probability that it comes from plant 2?

Solution

From Bayes' theorem given via Eq. (1.3),

$$P(A_2|B) = \frac{P(B|A_2) P(A_2)}{\sum_{i=1}^{3} P(B|A_i) P(A_i)} = \frac{0.05 \times 0.35}{0.029} = 0.603$$

1.3.3 Statistical Independence

Two random events A and B are independent if and only if

$$P(A \cap B) = P(A) P(B)$$

In what follows, we shall define random variables, and probability distribution and density functions.

1.4 RANDOM VARIABLES AND PROBABILITY DISTRIBUTION AND DENSITY FUNCTIONS

1.4.1 Random Variables

Let $\mathscr{E}$ be an experiment and S be the corresponding sample space. Then a random variable is a real function $X(\cdot)$ from S into the set of real numbers, i.e., for every $\xi \in S$, $X(\xi)$ is real.

The choice of the term "random variable" is not very appropriate because $X(\cdot)$ is a function, not a variable. However, we shall use the terminology in order to be consistent with the literature. In general, the random variables may be real or complex; however, unless specified otherwise, $X(\cdot)$ is assumed to be real. The random variable may be continuous or discrete.

Example 4

A fair coin is tossed three times. The sample space S is now considered to be:

$$S = \{HHH, HHT, HTH, HTT, THH, THT, TTH, TTT\}$$

where H denotes head and T denotes tail. Define $X(\cdot)$ = number of heads. Thus, $X(HHH) = 3$, $X(HHT) = 2$, etc. The random variable so defined is discrete.

1.4.2 Probability Distribution and Density Functions

Let $X(\cdot)$ be a continuous (piecewise continuous) random variable. Then the distribution function corresponding to $X(\cdot)$ is defined as:

$$F_X(\alpha) = P\{\xi \in S : X(\xi) \leq \alpha\} \qquad (1.4)$$

where α is a real number.

Before continuing the discussion, let us define the following notations:
(1) $[X \leq x] \triangleq \{\xi \in S: X(\xi) \leq x\}$
(2) $[X > x] \triangleq \{\xi \in S: X(\xi) > x\}$
(3) $[a < X < b] \triangleq \{\xi \in S: a < X(\xi) < b\}$

Thus, $F_X(\alpha)$ can now be written as:

$$F_X(\alpha) = P[X \leq \alpha] \qquad (1.5)$$

It is obvious that $F_X(\alpha)$ is a nondecreasing function.

Let us single out those random variables such that there exists a function $f_X(\cdot) \geq 0$, where

$$F_X(x) = \int_{-\infty}^{x} f_X(t)\, dt \tag{1.6}$$

The function $f_X(x)$ is called the probability density function (p.d.f.). If $f_X(x)$ is continuous (piecewise continuous), utilizing the Fundamental Theorem of Calculus, we obtain:

$$f_X(x) = \frac{dF_X(x)}{dx} \tag{1.7}$$

$f_X(x)$ is sometimes defined via Eq. (1.7).

It is also easy to verify the following properties:

(1) $P[a < X < b] = \int_a^b f_X(t)\, dt = F_X(b) - F_X(a)$

(2) $F_X(\infty) = \int_{-\infty}^{\infty} f_X(t)\, dt = 1$

(3) $F(-\infty) = 0$

(4) If $f_X(x)$ is continuous, then

$$P[x \leq X \leq x + \Delta x] = \int_x^{x+\Delta x} f_X(t)\, dt = \Delta x\, f_X(\xi)$$

where $\Delta x > 0$ and $x \leq \xi \leq x + \Delta x$ (using the Mean Value Theorem of Integrals).

(5) $P[X \geq x] = 1 - P[X \leq x] = 1 - F_X(x)$

(6) If $X(\cdot)$ is discrete, then $P(X_i) \geq 0$ and $\sum_{i=1}^{\infty} P(X_i) = 1$

Let us now define $F_X(\cdot)$ for the case where $X(\cdot)$ is a discrete random variable:

$$F_X(x) = P[X \leq x] = \sum_{x_i \leq x} P(x_i)$$

Henceforth, we shall drop the subscript X from $F_X(\cdot)$ and $f_X(\cdot)$ if there is no ambiguity about the random variable $X(\cdot)$.

Some examples of common continuous distributions are given below.

(1) *Uniform*

$$f_X(x) = \begin{cases} \dfrac{1}{b-a}, & a \leq x \leq b \\ 0, & \text{otherwise} \end{cases}$$

$$F_X(x) = \begin{cases} 0, & x < a \\ \dfrac{x-a}{b-a}, & a \leq x \leq b \\ 1, & x > b \end{cases}$$

(2) *Gaussian or Normal*

$$f_X(x) = \frac{1}{\sqrt{2\pi}\,\sigma} \exp\left[\frac{-(x-m)^2}{2\sigma^2}\right]$$

$$F_X(x) = \int_{-\infty}^{x} f_X(\alpha)\, d\alpha$$

where m and σ are parameters.

(3) *Rayleigh*

$$f_X(x) = \begin{cases} 0, & x < 0 \\ (x/a^2) \exp[-x^2/(2a^2)], & x \geq 0 \end{cases}$$

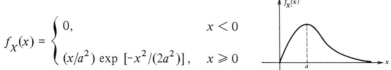

If there are two random variables $X_1(\cdot)$ and $X_2(\cdot)$ with possible outcomes x_1 and x_2, then a two-dimensional joint distribution function is defined as:

$$F_{X_1 X_2}(x_1, x_2) \triangleq P[X_1 \leq x_1 \text{ and } X_2 \leq x_2] \qquad (1.8)$$

Similar to the one-dimensional case, the two-dimensional probability density function $f_{X_1 X_2}(x_1, x_2)$ is a function such that:

$$f_{X_1 X_2}(x_1, x_2) \triangleq \frac{\partial^2 F_{X_1 X_2}(x_1, x_2)}{\partial x_1 \, \partial x_2} \qquad (1.9)$$

whenever $\partial^2 F / \partial x_1 \partial x_2$ exists. It can be easily be shown that:

$$F_{X_1 X_2}(x_1, x_2) = \int_{-\infty}^{x_1} \int_{-\infty}^{x_2} f_{X_1 X_2}(\alpha_1, \alpha_2) \, d\alpha_1 \, d\alpha_2 \qquad (1.10)$$

The following properties are true for joint distributions:

(1) $F_{X_1 X_2}(\infty, \infty) = 1$, $F_{X_1 X_2}(-\infty, -\infty) = 0$

(2) $F_{X_1 X_2}(x_1, x_2)$ is nondecreasing with respect to each argument

(3) $F_{X_1 X_2}(\infty, x_2) = F_{X_2}(x_2)$ and $F_{X_1 X_2}(x_1, \infty) = F_{X_1}(x_1)$

(4) $f_{X_1 X_2}(x_1, x_2) \geq 0$, for all x_1 and x_2

(5) $\iint_{-\infty}^{\infty} f_{X_1 X_2}(\alpha_1, \alpha_2) \, d\alpha_1 \, d\alpha_2 = 1$

The distribution and the probability density functions $F_{X_1}(x_1)$ and $f_{X_1}(x_1)$ are called marginal probability distribution and density functions (statistics), respectively, and that:

$$F_{X_1}(x_1) = F_{X_1 X_2}(x_1, \infty)$$

$$f_{X_1}(x_1) = \frac{\partial F_{X_1}(x_1)}{\partial x_1}$$

The marginal statistics $F_{X_2}(x_2)$ and $f_{X_2}(x_2)$ are defined in a similar manner.

Let A and B be events such that:
$$A = [X_1 \leq \alpha] \text{ and } B = [\beta_1 \leq X_2 \leq \beta_2]$$

Then from Eq. (1.1),

$$P(A|B) = \frac{P(A \cap B)}{P(B)} = \frac{\int_{-\infty}^{\alpha} \int_{\beta_1}^{\beta_2} f_{X_1 X_2}(x_1, x_2) \, dx_1 \, dx_2}{\int_{\beta_1}^{\beta_2} f_{X_2}(x_2) \, dx_2} \quad (1.11)$$

where $P(B)$ is assumed to be $\neq 0$.

Now, as $\beta_2 \to \beta_1 = \beta$,

$$F_{X_1}(\alpha | X_2 = \beta) = P(A|B) = \frac{\int_{-\infty}^{\alpha} f_{X_1 X_2}(x_1, \beta) \, dx_1}{f_{X_2}(\beta)} \quad (1.12)$$

The conditional p.d.f. $f_{X_1}(\alpha | X_2 = \beta)$ is given by:

$$f_{X_1}(\alpha | X_2 = \beta) = \frac{\partial F_{X_1}(\alpha | X_2 = \beta)}{\partial \alpha} \quad (1.13)$$

Utilizing Eq. (1.12) yields:

$$f_{X_1}(\alpha | X_2 = \beta) = \frac{f_{X_1 X_2}(\alpha, \beta)}{f_{X_2}(\beta)} \quad (1.14)$$

In a similar manner, we can show:

$$f_{X_2}(\beta | \alpha) = \frac{f_{X_1 X_2}(\alpha, \beta)}{f_{X_1}(\alpha)} \quad (1.15)$$

By combining the last two equations,

$$f_{X_1}(\alpha|\beta) = \frac{f_{X_2}(\beta|\alpha) f_{X_1}(\alpha)}{f_{X_2}(\beta)} \quad (1.16)$$

The last expression is called the Bayes' theorem for probability density functions and it is similar to the Bayes' theorem stated for the probability.

The conditional density concepts can easily be extended to the vector case.

1.5 FUNCTIONS OF RANDOM VARIABLES

For the sake of simplicity we shall discuss the function of a single random variable and then extend it to multivariables.

Let $X(\cdot)$ be a random variable and let $g(\cdot)$ be a real valued function such that

$$y = g(x)$$

and suppose $F_X(x)$ and $f_X(x)$ are given. Let us find $F_Y(y)$ and $f_y(y)$. We shall give the results via the following theorem.

Theorem 1

Let $g(x)$ be a piecewise continuously differentiable function and that for every y there exists m points $x_1, x_2, \ldots, x_m$ such that

$$y = g(x_k), \quad k = 1, 2, \ldots, m$$

and

$$g'(x_k) \neq 0, \quad k = 1, 2, \ldots, m$$

Then the following will hold:

$$f_Y(y) = \frac{f_X(x_1)}{|g'(x_1)|} + \ldots + \frac{f_X(x_m)}{|g'(x_m)|} \quad (1.17)$$

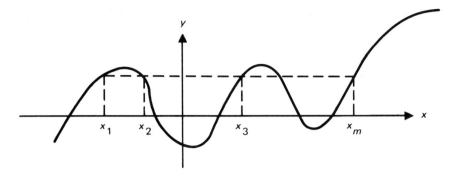

The proof is not given here. However, the proof can be constructed as the generalization of the case where $g(\cdot)$ is one-to-one and $g'(x) > 0 \; \forall^* \; x$ (or $g'(x) < 0$). For a proof, see references [1], [9], or [10].

Example 5

Let X and Y be random variables such that

$$Y = aX + b$$

where a and b are real constants. Assuming $F_X(x)$ and $f_X(x)$ are known, let us obtain $F_Y(y)$ and $f_Y(y)$.

Solution

$$F_Y(y) = P[Y \leq y] = P[aX + b \leq y]$$

$$= P\left[X \leq \frac{y-b}{a}\right] = F_X\left(\frac{y-b}{a}\right)$$

Now $f_Y(y)$ can be obtained via Eq. (1.17). Thus,

$$f_Y(y) = \frac{f_X(x)}{|g'(x)|} = \frac{f_X\left(\frac{y-b}{a}\right)}{|a|}$$

---

*$\forall$ means "for all."

Example 6

Let X and Y be random variables such that:

$$Y = g(X) = X^2$$

Obtain $F_Y(y)$ and $f_Y(y)$ assuming $F_X(x)$ and $f_X(x)$ are known.

Solution

$$F_Y(y) = P[Y \leq y] = P[X^2 \leq y] = P[-\sqrt{y} \leq X \leq \sqrt{y}] = F_X(\sqrt{y}) - F_X(-\sqrt{y})$$

If $y > 0$, $f_Y(y)$ can be calculated as:

$$f_Y(y) = \frac{f_X(x_1)}{|g'(x_1)|} + \frac{f_X(x_2)}{|g'(x_2)|} = \frac{f_X(-\sqrt{y})}{|2(-\sqrt{y})|} + \frac{f_X(\sqrt{y})}{|2(\sqrt{y})|}$$

Thus,

$$f_Y(y) = \begin{cases} \dfrac{1}{2\sqrt{y}} [f_X(-\sqrt{y}) + f_X(\sqrt{y})], & \text{if } y > 0 \\ \\ 0, & \text{otherwise} \end{cases} \quad (1.18)$$

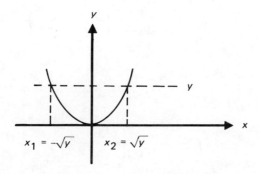

which completes the problem.

Let X and Y be random variables with the joint p.d.f. $f_{XY}(x,y)$ and let

$$z = g(x,y) \text{ and } w = h(x,y)$$

be real and continuous differentiable functions. We can obtain $f_{ZW}(z,w)$ in terms of $f_{XY}(x,y)$. For the sake of simplicity, let us assume that $g(x,y)$ and $h(x,y)$ are one-to-one functions. Then, it can be shown that:

$$f_{ZW}(z,w) = \frac{f_{XY}(x,y)}{|J(x,y)|}, \quad \text{assuming } J(x,y) \neq 0 \quad (1.19)$$

where x and y must be solved in terms of z and w, and $J(x,y)$ is given by:

$$J(x,y) = \begin{vmatrix} \frac{\partial g(x,y)}{\partial x} & \frac{\partial g(x,y)}{\partial y} \\ \frac{\partial h(x,y)}{\partial x} & \frac{\partial h(x,y)}{\partial y} \end{vmatrix} \quad (1.20)$$

If there are

$$(x_1, y_1), \ldots, (x_m, y_m)$$

ordered pairs such that

$$z = g(x_i, y_i) \text{ and } w = h(x_i, y_i), \quad i = 1, 2, \ldots, m$$

then Eq. (1.20) can be generalized by:

$$f_{ZW}(z,w) = \sum_{i=1}^{m} \frac{f_{XY}(x_i, y_i)}{|J(x_i, y_i)|}, \quad \text{assuming } J(x_i, y_i) \neq 0, \text{ for all } i \quad (1.21)$$

The result can be extended to the general case, where we are dealing with an n-random vector $\mathbf{X}$.

Let $\mathbf{X} = (X_1, \ldots, X_n)$ and $\mathbf{Y} = (Y_1, \ldots, Y_n)$ be random vectors such that:

$$\mathbf{Y} = h(\mathbf{X}) \tag{1.22}$$

and, for the sake of simplicity, assume h is one-to-one, i.e., invertible.

Let g be the inverse function given by:

$$\mathbf{X} = g(\mathbf{Y}) = g(h(\mathbf{X})) \tag{1.23}$$

Let A and B be events such that $B = [\mathbf{Y} \leq \mathbf{y}]$ and $A = [\mathbf{X} \leq g(\mathbf{y})]$. Remember that the notation $[\mathbf{Y} \leq \mathbf{y}]$ means $\{\xi \in S : y_i(\xi) \leq y_i \text{ for all } i = 1, 2, \ldots, m\}$. It is obvious that

$$F_{\mathbf{Y}}(\mathbf{y}) = F_{\mathbf{X}}(g(\mathbf{y}))$$

since they both represent the same probability. Thus,

$$\int_{-\infty}^{y} f_{\mathbf{Y}}(\alpha)\, d\alpha = \int_{-\infty}^{g(y)} f_{\mathbf{X}}(\beta)\, d\beta \tag{1.24}$$

The last integral is actually:

$$\int_{-\infty}^{y_1} \cdots \int_{-\infty}^{y_n} f_{Y_1 \ldots Y_n}(y_1, \ldots, y_n)\, dy_1 \cdots dy_n =$$

$$\int_{-\infty}^{x_1 = g_1(y_1, \ldots, y_n)} \cdots \int_{-\infty}^{x_n = g_n(y_1, \ldots, y_n)} f_{X_1 \ldots X_n}(\beta_1, \ldots, \beta_n)\, d\beta_1 \cdots d\beta_n$$

$$\tag{1.25}$$

If we differentiate Eq. (1.24) or (1.25) integrals with respect to each component of $\mathbf{y}$, we obtain:

$$f_{\mathbf{Y}}(\mathbf{y}) = f_{\mathbf{X}}(g(\mathbf{y})) \left| \frac{\partial g(\mathbf{y})}{\partial \mathbf{y}} \right| \tag{1.26}$$

where $\partial g(y)/\partial y$ is the determinant of the Jacobian:

$$\begin{bmatrix} \dfrac{\partial g_1}{\partial \beta_1} & \cdots & \dfrac{\partial g_1}{\partial \beta_n} \\ \vdots & & \vdots \\ \dfrac{\partial g_n}{\partial \beta_1} & \cdots & \dfrac{\partial g_n}{\partial \beta_n} \end{bmatrix}$$

Equation (1.25) can also be rewritten as (assuming $\partial g(y)/\partial y \neq 0$):

$$f_Y(y) = \frac{f_X(g(y))}{\left|\left(\dfrac{\partial g(y)}{\partial y}\right)^{-1}\right|} = \frac{f_X(g(y))}{|J(x)|} \tag{1.27}$$

where

$$J(x) = J(x_1, \ldots, x_n) = \begin{vmatrix} \dfrac{\partial h_1}{\partial x_1} & \cdots & \dfrac{\partial h_1}{\partial x_n} \\ \vdots & & \vdots \\ \dfrac{\partial h_n}{\partial x_1} & \cdots & \dfrac{\partial h_n}{\partial x_n} \end{vmatrix} = \left(\dfrac{\partial g(y)}{\partial y}\right)^{-1}$$

If h is not a one-to-one function, the result can be extended in a manner similar to Eq. (1.21).

1.6 SOME USEFUL DEFINITIONS AND CONCEPTS

Let X be a random variable and $g(\cdot)$ be a real function. Then the "expectation" or the "mean" of $g(x)$ is defined as the Stieltjes integral:

$$E[g(x)] = \int_{-\infty}^{\infty} g(x) \, dF_X(x) \tag{1.28}$$

If the reader is not familiar with the Stieltjes integral, then Eq. (1.28), when $F_X(x)$ is differentiable, would reduce to:

$$E[g(x)] = \int_{-\infty}^{\infty} g(x) f_X(x) \, dx \tag{1.29}$$

which is used in most engineering books.

The "variance" of X is denoted as σ_X^2 and is defined as:

$$\sigma_X^2 = E(X - m)^2 \tag{1.30}$$

where $m = EX$, and σ_X is called the "standard deviation." It can be shown that:

$$\sigma_X^2 = E(X^2) - m^2 \tag{1.31}$$

We shall also have the simple but useful inequalities:

$$P[\,|X| \geqslant K\,] \leqslant \frac{E\,|X|^n}{K^n}$$

and

$$P[\,|X - m| \geqslant K\sigma_X\,] \leqslant \frac{1}{K^2}$$

where K is a positive number and n is any integer such that $E[\,|X|^n\,] < \infty$.

If $\mathbf{X}(\cdot)$ is a random vector, then

$$\mathbf{X}(\xi) = (X_1(\xi), X_2(\xi), \ldots, X_n(\xi))$$

where $\xi \in S$. The case where $n = 2$ and $\mathbf{X}(\xi) = (X_1(\xi), X_2(\xi)) = (x_1, x_2) = x_1 + jx_2$ is defined as the complex random variable and it can be shown that:

$$E|X_1 X_2| \leqslant (E|X_1|^p)^{1/p} (E|X_2|^q)^{1/q} \qquad (1.32)$$

where p and q are greater than 1 and $(1/p) + (1/q) = 1$. The above equation is called the Hölder inequality.

For the special case, where $p = q = 2$, we get:

$$E|X_1 X_2| \leqslant (E|X_1|^2)^{1/2} (E|X_2|^2)^{1/2} \qquad (1.33)$$

Equation (1.33) is called the Schwarz inequality and will be used often.

1.6.1 Covariance and Correlation Coefficient

Let m_i and σ_i^2 be the mean and the variance of X_i, and let us define

$$\mu_{ij} = E[(X_i - m_i)(X_j - m_j)]$$

Then from the definition it is obvious that $\mu_{ii} = \sigma_i^2$, and, for $i \neq j$, we call μ_{ij} the covariance of X_i and X_j and ρ_{ij} defined by:

$$\rho_{ij} = \frac{\mu_{ij}}{\sigma_i \sigma_j} \qquad (1.34)$$

as the correlation coefficient between X_i and X_j. It can be checked that $-1 \leqslant \rho_{ij} \leqslant 1$ or, equivalently, $|\rho_{ij}| \leqslant 1$.

The matrix Λ_X is defined by:

$$\Lambda_X = \begin{bmatrix} \mu_{11} & \mu_{12} & \cdots & \mu_{1n} \\ \mu_{21} & \mu_{22} & \cdots & \mu_{2n} \\ \cdot & & & \\ \cdot & & & \\ \cdot & & & \\ \mu_{n1} & \mu_{n2} & \cdots & \mu_{nn} \end{bmatrix} \quad (1.35)$$

and is called the covariance matrix. Note that $\mu_{ij} = \mu_{ji}$; thus Λ_X is a symmetric matrix and, using the Schwarz inequality given by Eq. (1.33), we have:

$$|\mu_{ij}| \leq \sigma_i \sigma_j = |\mu_{ii}|^{1/2} |\mu_{jj}|^{1/2} \quad (1.36)$$

which verifies $|\rho_{ij}| \leq 1$. If $|\Lambda_X| \neq 0$ or, equivalently, the matrix Λ_X has the rank n, we say Λ_X is nonsingular.

1.6.2 Convergence

Let $X_1, X_2, \ldots, X_n, \ldots$ and X be random variables defined from $S \to R$. Then the set $A = \{\xi : X_n(\xi) \to X(\xi)\}$ is an event (that is, $A \subset S$). Thus the probability that X_n converges to X is defined.

There are several criteria of convergence. The following modes are defined for both real and complex valued random variables as $n \to \infty$:

(1) X_n converges in probability (or P-measure) to X, if for any given $\epsilon > 0$, $P(|X_n - X| > \epsilon) \to 0$ (or $\lim P(|X_n - X| > \epsilon) = 0$ as $n \to \infty$).

(2) X_n converges in quadratic mean or mean square (m.s.) to X if $E(|X_n - X|^2) \to 0$.

(3) X_n converges with probability one or "almost everywhere" to X if $P(X_n \to X) = 1$, or, equivalently, $P(X_n \not\to X) = 0$.

Example 7

Consider the uniform p.d.f. $f_X(x)$ given via the figure below.

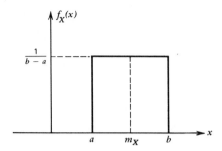

Calculate the mean and the variance of X.

Solution

$$m_X = \int_{-\infty}^{\infty} x f_X(x) dx = \frac{1}{b-a} \int_a^b x\, dx = \frac{b+a}{2}$$

Since the variance $\sigma_X^2 = E(X^2) - m_X^2$, we need to obtain $E(X^2)$.

$$E(X^2) = \int_{-\infty}^{\infty} x^2 f_X(x) dx = \frac{1}{b-a} \int_a^b x^2\, dx = \frac{1}{3}(b^2 + ab + a^2)$$

Then, it is obvious that

$$\sigma_X^2 = E(X^2) - m_X^2 = \frac{1}{12}(b-a)^2$$

Consequently, the standard deviation σ_X is given by:

$$\sigma_X = \frac{1}{2\sqrt{3}}(b-a)$$

Example 8

Assume we have the two-dimensional random variable (X, Y) which is distributed uniformly over a region D given by:

$$D = \{(x,y) : 0 < x < y < 1\}$$

Also see the corresponding figure given below.

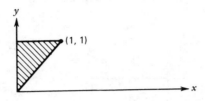

Determine the correlation coefficient.

Solution

Let ρ denote the correlation coefficient, thus ρ is given by:

$$\rho = \frac{E[(X - m_X)(Y - m_Y)]}{\sigma_X \sigma_Y} = \frac{E(XY) - m_X m_Y}{\sigma_X \sigma_Y}$$

Since the area given by D is $1/2$, the joint p.d.f. $f_{XY}(x, y)$ is given by:

$$f_{XY}(x, y) = \begin{cases} 2, & \text{if } (x, y) \in D \\ 0, & \text{otherwise} \end{cases}$$

In order to obtain ρ, we need to obtain $E(XY)$, m_X, m_Y, σ_X, and σ_Y. The last four quantities require the knowledge of $f_X(x)$ and $f_Y(y)$, which for $(x, y) \in D$ are given below

$$f_X(x) = \int_{-\infty}^{\infty} f_{XY}(x, y) dy = \int_x^1 2 \, dy = 1(1 - x)$$

and

$$f_Y(y) = \int_{-\infty}^{\infty} f_{XY}(x, y) dx = \int_0^y 2 \, dy = 2y$$

We can now obtain $m_X, m_Y, \sigma_X, \sigma_Y$, and $E(XY)$ as follows:

$$m_X = \int_0^1 2x(1-x)dx = \frac{1}{3} \text{ and } m_Y = \int_0^1 2y^2\, dy = \frac{2}{3}$$

$$\sigma_X^2 = E(X^2) - m_X^2 = \int_0^1 2x^2(1-x)dx - \frac{1}{9} = \frac{1}{18}$$

$$\sigma_Y^2 = E(Y^2) - m_Y^2 = \int_0^1 2y^3\, dy - \frac{4}{9} = \frac{1}{18}$$

$$E(XY) = \int_0^1 \int_0^y 2xy\, dx\, dy = \frac{1}{4}$$

Hence, ρ is given by:

$$\rho = \frac{E(XY) - m_X m_Y}{\sigma_X \sigma_Y} = \frac{1/4 - (1/3)(2/3)}{\sqrt{1/18}\sqrt{1/18}} = \frac{1}{2}$$

Example 9

Let X and Y be two independent random variables such that $f_X(x) = \exp(-x)$ and $f_Y(y) = 2\exp(-2y)$, where both x and $y \geq 0$. Define $z = x/y$ and obtain $f_Z(z)$.

Solution

Because X and Y are independent, $f_{XY}(x,y) = f_X(x)f_Y(y)$, therefore:

$$f_{XY}(x,y) = 2\exp(-x + 2y) \text{ whenever } x \text{ and } y \geq 0$$

It is left as an exercise to verify that

$$f_Z(z) = \int_{-\infty}^{\infty} |y| f_{XY}(x,y) dx\, dy = 2\int_{-\infty}^{\infty} |y| \exp[-(z+2)y]\, dy$$

whenever x and $y \geqslant 0$. One can now obtain that:

$$f_Z(z) = 2 \int_0^\infty y \exp[-(z+2)y]\, dy = \frac{2}{(z+2)^2} \quad \text{for } z \geqslant 0$$

Example 10

A one-mile runner starts running the mile, but he changes his mind (between the start and the finish line) and decides to walk the rest of the way. Assuming he walks on the same path and gets to the nearest end point (start or finish), what is the expected value of the distance he walks?

Solution

Let the start and the finish of the one mile be given by 0 or 1, respectively (see the figure below).

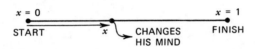

Let y be the walking distance from the point x to the start or finish, whichever is closer. It is obvious from the figure that $0 \leqslant x \leqslant 1$. Since $f_X(x)$ is uniform, then

$$f_X(x) = \begin{cases} 1, & 0 \leqslant x \leqslant 1 \\ 0, & \text{otherwise} \end{cases}$$

Thus, y becomes a function of x, say $y = g(x)$. Then

$$y = g(x) = \begin{cases} x, & \text{if } x < 1/2 \\ 1-x, & \text{if } x > 1/2 \end{cases}$$

and the relation between x and y is given by the figure below.

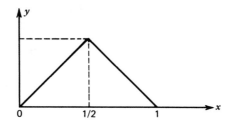

Now we must obtain $E(Y)$, which is

$$E(Y) = \int_{-\infty}^{\infty} y f_Y(y) dy = \int_{-\infty}^{\infty} g(x) f_X(x) dx$$

$$= \int_0^{1/2} x\, dx + \int_{1/2}^1 (1-x) dx = \frac{1}{4}$$

This concludes the problem.

Example 11

Let $f_{XY}(x, y)$ be given by

$$f_{XY}(x, y) = \begin{cases} \alpha(x + y), & \text{if } 0 \leq x \text{ and } y \leq 1 \\ 0, & \text{otherwise} \end{cases}$$

(a) Find α so that $f_{XY}(x, y)$ is a p.d.f.
(b) Show whether or not X and Y are independent.
(c) Obtain $E(Y|X)$ as a function of x. Show whether or not this function is defined for x outside the interval $[0, 1]$.
(d) If we define a random variable $Z = E[Y|X]$, then obtain $f_Z(z)$.

Solution

(a) From

$$\int_{-\infty}^{\infty} \int f_{XY}(x, y) dx\, dy = \int_0^1 \int_0^1 \alpha(x + y) dx\, dy = 1$$

we get:

$$\alpha \int_0^1 \int_0^1 (x+y)dx\, dy = \alpha \int_0^1 (1/2 + y)dy = \alpha = 1$$

Therefore, $\alpha = 1$.

(b) For X and Y to be independent we must have $f_{XY}(x,y) = f_X(x)f_Y(y)$. The equality will not hold, since for $0 \leqslant x$ and $y \leqslant 1$, we have

$$f_X(X) = x + 1/2 \quad \text{and} \quad f_Y(y) = y + 1/2$$

with $f_{XY}(x,y) \neq f_X(x)f_Y(y)$.

(c) $E(Y|X) = \int_{-\infty}^{\infty} y f_X(Y|X) dy$

Thus, we must first obtain $f_{XY}(x,y)$. For $f_X(x) \neq 0$, we know:

$$f(Y|X) = \frac{f_{XY}(x,y)}{f_X(x)} = \begin{cases} \dfrac{x+y}{x+1/2}, & \text{if } 0 \leqslant x \text{ and } y \leqslant 1 \\ 0, & \text{otherwise} \end{cases}$$

Now for $0 \leqslant x \leqslant 1$

$$E(Y|X) = \int_0^1 y\left(\frac{x+y}{x+1/2}\right) dy = \frac{x/2 + 1/3}{x + 1/2}$$

and it is not defined otherwise.

(d) For $z = (x/2 + 1/3)/(x + 1/2)$ and x defined in the closed interval $[0, 1]$, we shall obtain $5/9 \leqslant z \leqslant 2/3$, and

$$f_Z(z) = f_X(x) \left|\frac{dx}{dz}\right|$$

We can solve for x in terms of z and dx/dy can be found as:

$$x = \frac{1/3 - z/2}{z - 1/2} \quad \text{and} \quad \frac{dx}{dy} = \frac{-1}{12(z - 1/2)^2}$$

Consequently,

$$f_Z(z) = \begin{cases} \dfrac{1}{144(z - 1/2)^3}, & \text{if } 5/9 \leq z \leq 2/3 \\ 0, & \text{elsewhere} \end{cases}$$

1.7 NORMAL DISTRIBUTIONS AND CHARACTERISTIC EQUATIONS

The most important distribution is the normal distribution. The normal p.d.f. $f_X(x)$ is defined as:

$$f_X(x) = \frac{1}{\sqrt{2\pi}\,\sigma} \exp\left[-\frac{1}{2}\left(\frac{x - m}{\sigma}\right)^2\right] \qquad (1.37)$$

where X is a random variable (one-dimensional).

The error function erf(x) is defined as:

$$\text{erf}(x) = \frac{1}{\sqrt{2\pi}} \int_0^x \exp\left[-\frac{y^2}{2}\right] dy \qquad (1.38)$$

It can be easily verified that:

$$F_X(x) = \frac{1}{2} + \text{erf}\left[\frac{x - m}{\sigma}\right] = \frac{1}{2} + \frac{1}{\sqrt{2\pi}} \int_0^{(x-m)/\sigma} \exp\left[-\frac{y^2}{2}\right] dy \qquad (1.39)$$

Note that if we take the derivative of $F(x)$ we get $f(x)$, i.e.,

$$f_X(x) = 0 + \frac{1}{\sqrt{2\pi}} \exp\left[-\frac{1}{2}\left(\frac{x-m}{\sigma}\right)^2\right] \left(\frac{d}{dx}\left(\frac{x-m}{\sigma}\right)\right)$$

$$= \frac{1}{\sqrt{2\pi}\,\sigma} \exp\left[-\frac{1}{2}\left(\frac{x-m}{\sigma}\right)^2\right]$$

as asserted.

Note that in the above equation we have used the Fundamental Theorem of Calculus, which states: If

$$G(x) = \int_{h_1(x)}^{h_2(x)} g(y)\,dy$$

where h_1 and h_2 are differentiable and g is continuous, then

$$\frac{dG(x)}{dx} = g(h_2(x))\frac{dh_2}{dx} - g(h_1(x))\frac{dh_1}{dx} \qquad (1.40)$$

From the above equation we get:

$$F_X(x_2) - F_X(x_1) = \mathrm{erf}\left[\frac{x_2 - m}{\sigma}\right] - \mathrm{erf}\left[\frac{x_1 - m}{\sigma}\right]$$

It can be verified that for the normal distribution the p.d.f. is symmetric about the mean m and

$$E\{|X|^n\} = \begin{cases} 0, & n = \text{odd} \\ 1 \cdot 3 \cdot 5 \cdots (2k-1)\sigma^{2k}, & n = 2k \text{ (even)} \end{cases}$$

Also, it can be shown that if X_1 and X_2 are independent normal random variables, with respective (m_1, σ_1) and (m_2, σ_2), then their sum $X = x_1 + x_2$ is also normal with mean $m = m_1 + m_2$ and variance $\sigma^2 = \sigma_1^2 + \sigma_2^2$. Thus, the summation of independent normal random variables produces a new normal random variable. However, the "Central Limit Theorem" states (under fairly wide conditions) that the sum of a large number of independent random variables is approximately normally distributed, even though each individual random variable may not be normal.

Using the above results, one can verify that:

$$P[|x - m| < k\sigma] = 2 \operatorname{erf}(k)$$

where $k > 0$.

The verification is simple since

$$P[|x - m| < k\sigma] = P[-k\sigma + m < x < k\sigma + m] = F_X(k\sigma + m) - F_X(-k\sigma + m)$$

$$= \operatorname{erf}(k) - \operatorname{erf}(-k) = 2 \operatorname{erf}(k)$$

Using Eq. (1.38), $P[|x - m| < k\sigma]$ can be obtained for any k. Let us list $P[|x - m| < k\sigma]$ for different values of k.

$$P[|x - m| < k\sigma] = \begin{cases} 0.07966, & k = 0.1 \\ 0.38292, & k = 0.5 \\ 0.51607, & k = 0.7 \\ 0.63188, & k = 0.9 \\ 0.68269, & k = 1.0 \\ 0.86638, & k = 1.5 \\ 0.95449, & k = 2.0 \\ 0.99730, & k = 3.0 \\ 0.99993, & k = 4.0 \\ 0.99999, & k = 5 \end{cases}$$

The following figure shows a normal p.d.f. $f_X(x)$, where the shaded area corresponds to $P[|x - m| < k\sigma]$.

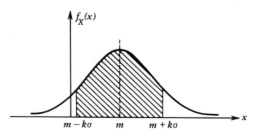

1.7.1 The Vector Case

Let $\mathbf{X} = (X_1, X_2, \ldots, X_n)^T$, where T is the transpose, be a normally distributed random vector; thus,

$$f_{\mathbf{X}}(x_1, x_2, \ldots, x_n) = \frac{1}{(2\pi)^{n/2} \sqrt{|\Lambda|}} \exp\left\{-\frac{1}{2}(\mathbf{x} - \mathbf{m})^T \Lambda^{-1}(\mathbf{x} - \mathbf{m})\right\}$$

(1.41)

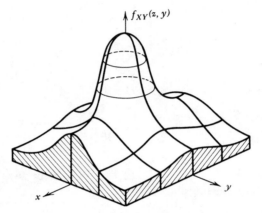

where Λ is the covariance of $\mathbf{X}$, i.e., $\Lambda \triangleq E[(\mathbf{x} - \mathbf{m})(\mathbf{x} - \mathbf{m})^T]$, $|\Lambda|$ is the determinant of Λ, and

$$\mathbf{m} = \begin{bmatrix} m_1 \\ . \\ . \\ . \\ m_n \end{bmatrix} = E(\mathbf{X})$$

It can be shown that Λ can also be written as

$$\Lambda = E(\mathbf{X}\mathbf{X}^T) - \mathbf{m}\mathbf{m}^T$$

Notationally we can write $f_{\mathbf{X}}(\mathbf{x}) = G(\mathbf{x}, \mathbf{m}, \Lambda)$, which means the Gaussian density of $\mathbf{X}$ has the mean $\mathbf{m}$ and the covariance Λ.

In order to derive some important properties in the normal random vectors, we need some basic definitions.

1.8 THE CHARACTERISTIC FUNCTION

Recalling from the one-dimensional random variable, let X be a (one-dimensional) random variable. Then the characteristic function of X is defined as:

$$C(u) = E[\exp(juX)] = \int_{-\infty}^{\infty} \exp(jux) f_X(x)\, dx \qquad (1.42)$$

It is seen that the characteristic function is the Fourier transform of $f_X(x)$; however, the positive sign in the exponent simply means that we must use the negative sign in finding the inverse. Thus, the density function $f_X(x)$ can be obtained from (using the Fourier transform pair):

$$f_X(x) = \frac{1}{2\pi} \int_{-\infty}^{\infty} C(u) \exp(-jux)\, du \qquad (1.43)$$

For a discussion of the Fourier transforms, see Appendix C.

It can be shown that

$$C(u) = \sum_{k=-\infty}^{k=\infty} \frac{(ju)^k}{k!} m_k$$

using Eq. (1.42) where

$$m_k = \int_{-\infty}^{\infty} x^k f_X(x)\, dx$$

and making use of

$$m_k = (-j)^k \left. \frac{d^k C(u)}{du^k} \right|_{u=0}$$

The most useful property of the characteristic function is that it relates the sum of independent random variables. It is also used to simplify calculations.

1.9 DEFINITION EXTENDED TO RANDOM VECTORS

The characteristic function of a random variable $\mathbf{X} = (X_1, \ldots, X_n)^T$ is defined as:

$$C(\mathbf{u}) = C(u_1, \ldots, u_n) = E[\exp(j\mathbf{u}^T\mathbf{X})]$$

$$= \int\int \cdots \int_{-\infty}^{\infty} \exp(j\mathbf{u}^T\mathbf{x}) f_\mathbf{X}(x_1, \ldots, x_n) dx_1 dx_2 \cdots dx_n$$

(1.44)

Now let us apply the definition given by Eq. (1.44) to the Gaussian random vector $\mathbf{X} = (X_1, \ldots, X_n)^T$ and make the following claim:

Theorem 2

The characteristic function of the random vector $\mathbf{X}$ is given by:

$$C(\mathbf{u}) = \exp\left[j\mathbf{u}^T\mathbf{m} - \frac{1}{2}\mathbf{u}^T \Lambda \mathbf{u}\right]$$

Proof

Left as an exercise.

Theorem 3

If two normal vectors $\mathbf{X}$ and $\mathbf{Y}$ are Gaussian with respective means (vectors) $\mathbf{m_X}$ and $\mathbf{m_Y}$ and are also uncorrelated, then they are statistically independent.

Proof

Let $\mathbf{X}$ be n-dimensional and $\mathbf{Y}$ be m-dimensional with respective covariances $\Lambda_\mathbf{X}$ and $\Lambda_\mathbf{Y}$.

Define a vector

$$Z = \begin{bmatrix} X_1 \\ X_2 \\ \vdots \\ X_n \\ \hline Y_1 \\ Y_2 \\ \vdots \\ Y_m \end{bmatrix} \triangleq \begin{bmatrix} X \\ \hline Y \end{bmatrix}$$

Define Λ_{XY} (cross-covariance):

$$\Lambda_{XY} = E[(X - m_X)(Y - m_Y)^T]$$

Before proving the assertion, observe that $\Lambda_{XY}^T = E[(Y - m_Y)(X - m_X)^T] = \Lambda_{YX}$. Let us now calculate Λ_Z:

$$\Lambda_Z = E[Z - m_Z)(Z - m_Z)^T] = E\left\{\begin{bmatrix} X - m_X \\ \hline Y - m_Y \end{bmatrix}\begin{bmatrix} X - m_X \\ \hline Y - m_Y \end{bmatrix}^T\right\} = \begin{bmatrix} \Lambda_X & \Lambda_{XY} \\ \hline \Lambda_{YX} & \Lambda_Y \end{bmatrix}$$

Then

$$f_{XY}(x,y) = f(x_1, \ldots, x_n, y_1, \ldots, y_m) = G\left(\begin{bmatrix} x \\ y \end{bmatrix}, \begin{bmatrix} m_X \\ m_Y \end{bmatrix}, \begin{bmatrix} \Lambda_X & \Lambda_{XY} \\ \Lambda_{YX} & \Lambda_Y \end{bmatrix}\right)$$

$$= \left(\frac{1}{(2\pi)^{\frac{m+n}{2}} \sqrt{\left|\begin{matrix} \Lambda_X & \Lambda_{XY} \\ \Lambda_{YX} & \Lambda_Y \end{matrix}\right|}}\right) \cdot$$

$$\exp\left(-\frac{1}{2}\begin{bmatrix} X - m_X \\ Y - m_Y \end{bmatrix}^T \begin{bmatrix} \Lambda_X & \Lambda_{XY} \\ \Lambda_{YX} & \Lambda_Y \end{bmatrix}^{-1} \begin{bmatrix} X - m_X \\ Y - m_Y \end{bmatrix}\right)$$

If **X** and **Y** are uncorrelated, then $\Lambda_{XY} = 0$; hence

$$\left|\begin{matrix} \Lambda_X & \Lambda_{XY} \\ \Lambda_{YX} & \Lambda_Y \end{matrix}\right| \triangleq \left[\begin{matrix} \Lambda_X & \Lambda_{XY} \\ \Lambda_{YX} & \Lambda_Y \end{matrix}\right] = \left|\begin{matrix} \Lambda_X & 0 \\ 0 & \Lambda_Y \end{matrix}\right| = (\det \Lambda_X)(\det \Lambda_Y)$$

This substituted in $f(x,y)$ yields: $f(x,y) = f_X(x) f_Y(y)$. Done!

Theorem 4

If **X** and **Y** are specified as in Theorem 3, then we claim:

$$E(X|Y) = E((x_1, \ldots, x_n) \mid (y_1, \ldots, y_m)) = m_X + \Lambda_{XY} \Lambda_Y^{-1}(Y - m_Y)$$

and the conditional covariance matrix $\Lambda_{X|Y}$ is defined by:

$$\Lambda_{X|Y} = \{E(X - E(X|Y))(X - E(X|Y))^T\} = \Lambda_X - \Lambda_{XY} \Lambda_Y^{-1} \Lambda_{YX}$$

The proof is simple but lengthy and has been omitted.

EXERCISES

1.1 An urn contains 4 green and 6 blue marbles. Two marbles are drawn out together. One of them is tested and found to be blue. Find the probability that the other one is also blue.

1.2 Let A and B be independent events associated with an experiment. If the probability that A or B occurs is 0.7, while the probability of occurrence of A is 0.3, determine the probability of occurrence of B.

1.3 Three dice are thrown. Find the probabilities of the events of obtaining the sum of 10, 11, and 12 points.

1.4 A continuous random variable X has the distribution function:

$$F_X(x) = \begin{cases} 1 - (1 + \alpha x) \exp(-\alpha x), & \text{if } x > 0 \\ 0, & \text{if } x \leq 0 \end{cases}$$

(a) Find the characteristic function.

(b) Find the mean and the standard deviation.

1.5 Let the joint probability density function of the random vector (X, Y) be given by:

$$f_{XY}(x,y) = \begin{cases} xy \exp[-(x^2 + y^2)/2], & \text{if } x \text{ and } y \geq 0 \\ 0, & \text{otherwise} \end{cases}$$

(a) Find $f_X(x), f_Y(y), f(x|y),$ and $f(y|x)$.

(b) Are the random variables X and Y independent?

1.6 In the previous problem, if in addition we have the random variables Z and W given by:

(a) $Z = aX + bY$, $\quad W = cX + dY$

(b) $Z = YX^2\, U(X),\ W = XY^2\, U(Y)$

where $U(\cdot)$ is a unit step function, find $f_{ZW}(z, w)$.

1.7 Find the probability density functions of Z and W. Given:

$$f_{XY}(x,y) = 2 \exp - [x^2 + 2xy + 2y^2]$$

(a) Determine the mean and the variance of the random variable $Z = XY$.

(b) Determine the mean and the variance of the random variable $W = X^2 + Y$.

1.8 If X and Y are independent random variables such that:

$$f_X(x) = \begin{cases} \dfrac{1}{\pi} \dfrac{1}{(1-x^2)^{1/2}}, & \text{if } |x| < 1 \\ 0 & \text{otherwise} \end{cases}$$

$$f_Y(y) = \dfrac{y}{k^2} \exp[-y^2/2k^2]\, U(y)$$

where $U(y)$ is a unit step function. Show that the random variable $W = XY$ is normal with mean zero and variance k^2.

1.9 If in a vector case of a normal random vector, $n = 2$, $m_1 = m_2 = 0$ and $\mu_{11} = \mu_{22} = 1$, show that:

$$f(x_1, x_2, \rho) = \dfrac{1}{2\pi(1-\rho^2)^{1/2}} \exp\left[-\dfrac{x_1^2 + x_2^2 - 2\rho x_1 x_2}{2(1-\rho^2)}\right]$$

where $\rho = \mu_{21} = \mu_{12}$.

1.10

(a) If A is an $m \times n$ matrix such that

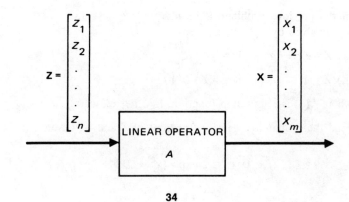

show that if **X** is normal so is **Z**. Use the property of characteristic equations given by Theorem 2. That is, show that the characteristic of **Z** is:

$$C(\mathbf{t}; \mathbf{m}_Z, \Lambda_Z) = \exp\left[j(\mathbf{t}^T \mathbf{m}_Z) - \frac{1}{2}(\mathbf{t}^T \Lambda_Z \mathbf{t})\right]$$

where $\mathbf{t} = \begin{bmatrix} t_1 \\ t_2 \\ \cdot \\ \cdot \\ \cdot \\ t_n \end{bmatrix}$

(b) Show that $\Lambda_X = A\, \Lambda_Z A^T$.

1.11 Two people are to meet between 12 noon and 1 p.m. They agree not to wait more than 15 minutes for each other. Assume each person arrives independently at random times between those hours. Find the conditional probability that the two meet, with the assumption that one of the people has arrived at 12:30 p.m.

1.12 Given $f_{XY}(x,y) = 1$ for $0 \leqslant x$ and $y \leqslant 1$ and zero otherwise with $z = x + y$, $w = x^2$, and $v = wz$. Find $E[V|Z]$ as a function of z.

1.13 Let X and Y be zero mean normal random variables with $\sigma_X = \sigma_Y = \sigma$. Let $f_{XY}(x, y) = f_X(x)f_Y(y)$ and $z = \sqrt{x^2 + y^2}$. Show that Z has a p.d.f. which is Rayleigh and its mean is non-zero.

CHAPTER 2
STOCHASTIC PROCESSES

2.1 INTRODUCTION

Very often we are interested in observations that are made over a period of time and that are affected by random chance. This situation is termed a stochastic process and is defined below.

2.2 DEFINITIONS AND EXAMPLES

Definition 1

A stochastic process $X(t,\omega)$ is a function of two variables, where ω is an element of the sample space and t is a parameter (time) which belongs to a set T (time interval).

Definition 2

For every $\omega_0 \in S$ (sample space), the function $X(t,\omega_0)$ is called a *sample function* of the process.

The process $X(t,\omega)$, in general, can be complex, but, without any loss of generality, we shall discuss $X(t,\omega)$ when it is real. Thus, to each sample point $\omega \in S$ (sample space), we are assigning a waveform X_t, which is the function of t (time) such that:

$$X_t: \omega \to X(t,\omega)$$

Hence, each sample space will have a collection of waveforms, each assigned to a member $\omega \in S$. The collection of all of these waveforms (as many as the cardinality of S) is called an ensemble. Thus, each individual member of the ensemble is a sample function.

Example 1

Assume that we toss a coin twice in succession. Then, our sample space S is the collection of four outcomes:

$$S = \{\underbrace{HH}_{\omega_1}, \underbrace{TT}_{\omega_2}, \underbrace{HT}_{\omega_3}, \underbrace{TH}_{\omega_4}\}$$

There exists four sample points ω_1, ω_2, ω_3, and ω_4. The probability of each occurrence is 1/4 (the coin is a fair one).

Let us now define a function $X_t(\cdot): S \to R$ such that:

$$X_t(\omega_k) = X(t,\omega_k) = \sin kt$$

Thus, the ensemble consists of four elements (as many as the cardinality of S, which is 4). Let us denote the ensemble by $\mathcal{E}$. Thus,

$$\mathcal{E} = \{\sin t, \sin 2t, \sin 3t, \sin 4t\}$$

and the probability assigned to each waveform is also 1/4.

Remark 1. The cardinality (number of sample points) corresponding to the sample space S may be finite, numberably infinite or dense.

Remark 2. For a stochastic process $X_t(\omega)$ or $X(t,\omega)$ is an appropriate designation. However, in common practice the process is represented by $X(t)$, which actually means $X(t,\omega)$.

2.2.1 More Words About X(t)

The notation of $X(t,\omega)$ may be better understood by the physical phenomenon. Consider a system such as a radar antenna receiver. Suppose the noise signal at the output is of interest. Each time we turn on the system, it will yield a different noise waveform. The collection of all of the noise waveforms is the ensemble of this process (see figure below).

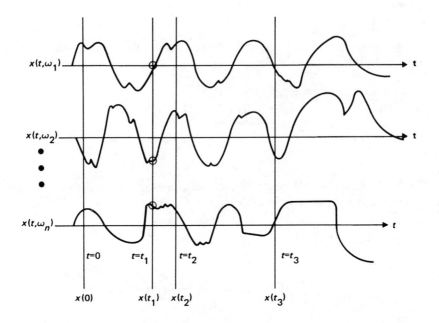

It is important to mention that each sample function (waveform) is assigned to a single point $\omega \in S$. Thus, after ω is specified, the waveform is deterministic (not random). The randomness is associated with each sample being chosen (occurrence of a sample).

Example 2

Suppose a receiver (antenna) detects signals of the form:

$$X(t) = a \cos(\omega t + \Theta)$$

where **a** (amplitude) and Θ are both random. Suppose by some sort of practical experience we know the distribution functions of Θ and **a** (for example, Θ or **a** could be Poisson, Gaussian, uniform, or any other probability density function).

Let us assume **a** is Gaussian and Θ is uniform over the open interval $(0, 2\pi]$, Then,

$$f_a(\beta) = \frac{1}{\sigma \sqrt{2\pi}} \exp\left[-\frac{(\beta - m)^2}{2\sigma^2}\right]$$

and

$$f_\Theta(\theta) = \begin{cases} \dfrac{1}{2\pi}, & \theta \in [0, 2\pi] \\ 0, & \text{elsewhere} \end{cases}$$

Corresponding to each sample function, **a** and Θ are assumed to be constant, but they definitely vary from one sample function to the other.

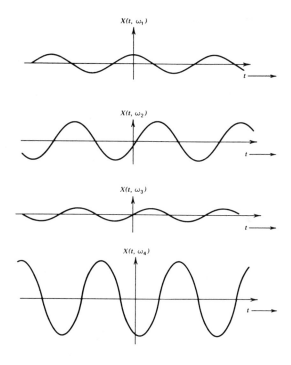

Example 3

Consider

$$X(t) = \mathbf{a}t + b$$

where **a** is a random variable, and b is a constant.

Remark 3. For the one-dimensional case $X(t, \omega)$ becomes a random variable for each fixed $t = t_1$ since $X(t_1, \omega)$ becomes a function of ω only, i.e.,

$$X(t_1, \omega): S \to R$$

which is the definition of the random variable.

Remark 4. Remember that we use the notation $X(t_1, \omega)$ (or $X_{t_1}(\omega)$) by either X_1 or $X(t_1)$.

2.3 FIRST-ORDER STATISTICS

The distribution of a real process $X(t)$ for a fixed $t = t_1$ is defined:

$$F_X(x, t_1) = P\{X(t_1) \leq x\} \tag{2.1}$$

Remember: $\{X(t_1) \leq x\} = \{\omega \in S : X(t_1, \omega) \leq x\}$.

Definition 3

The first-order statistics are those items of information that can be completely determined from $F_X(x, t)$, such as $f_X(x, t)$, $m(t) = EX(t)$ or $E[X(t)]^2$, $\sigma^2_{X(t)}$, etc.

Definition 4

A nonnegative function $f_X(x, t) \geq 0$, such that

$$F_X(x, t) = \int_{-\infty}^{x} f_X(x, t)\, dx \tag{2.2}$$

is called the probability density function (p.d.f.). If $F_X(x; t)$ is differentiable, then, from Eq. (2.2):

$$\frac{\partial F_X(x, t)}{\partial x} = f_X(x, t) \tag{2.3}$$

Note that condition (2.2) is a weaker condition than that of (2.3), because $f(x, t)$ may exist even though $F_X(x, t)$ may not be differentiable.

Note that:

$$E[X(t)] = \int_{-\infty}^{\infty} x f_X(x, t)\, dx$$

will be denoted as either $m(t)$ or $\eta(t)$ in what follows.

Example 4

Let us continue example 3, $X(t) = \mathbf{a}t + b$, where $t > 0$, b is a constant, and $\mathbf{a}$ is a Gaussian random variable:

$$f_\mathbf{a}(\alpha) = \frac{1}{\sqrt{2\pi}} \exp\left[-\frac{\alpha^2}{2}\right] \tag{2.4}$$

Find the first-order p.d.f. $f_X(x, t)$.

Solution

From $X(t) = \mathbf{a}t + b$, we get $\mathbf{a} = (1/t)(X - b)$. We know:

$$f_X(x, t) = \frac{f_\mathbf{a}(\alpha)}{\left|\dfrac{dx}{d\alpha}\right|}$$

Now $dx/d\alpha = t$; since $t > 0$, we have

$$\left|\frac{dx}{d\alpha}\right| = t$$

and

$$\alpha = \frac{1}{t}(x - b)$$

Hence,

$$f_X(x, t) = \frac{\exp\left[-\dfrac{\alpha^2}{2}\right]}{\sqrt{2\pi}\, t} = \frac{1}{t\sqrt{2\pi}} \left[\exp\left(-\frac{(x - b)^2}{2t^2}\right)\right]$$

Important Reminder. From now on, we shall drop the subscript X from $F_X(x, t)$ and $f_X(x, t)$ whenever it is appropriate.

Example 5

Obtain the mean and the variance of $X(t)$.

Solution

$$EX(t) = m(t) = tE(a) + E(b) = t \cdot 0 + b = b$$

Note: From Eq. (2.4) it is obvious that $E(a) = 0$, $\sigma^2 = 1 = E(a^2)$.

Since

$$\sigma^2_{X(t)} = E[X^2(t)] - E^2[X(t)]$$

then we must calculate $E[X^2(t)]$:

$$E[X(t)^2] = E[(at+b)^2] = E[t^2 a^2 + b^2 + 2tab]$$

$$= t^2 E(a^2) + b^2 + 2tb E(a) = t^2(1) + b^2 = t^2 + b^2$$

Hence,

$$\sigma^2_X = (t^2 + b^2) - E^2(X) = (t^2 + b^2) - b^2 = t^2$$

Remark 5. Regardless of the parameter t, the mean of $X(t)$ is b; however, both $E(X^2(t))$ and $\sigma^2_{X(t)}$ are dependent on t.

Example 6

Consider the random process $X(t)$ given by:

$$X(t) = A \cos(\omega_0 t + \Theta)$$

where Θ is a random variable which is uniformly distributed over $[0, 2\pi]$ and the amplitude A is constant.

Obtain the following first-order statistics:

(a) Probability density function

(b) $m(t)$

(c) The variance of X

Solution

(a) We can consider the sample function x to be

$$x = A\cos(\omega_0 t + \theta)$$

where x and θ denote the parameters (possible values of a random variable X and Θ, respectively). Since Θ is uniformly distributed, we get:

$$f_\Theta(\theta) = \begin{cases} \dfrac{1}{2\pi}, & \theta \in [0, 2\pi] \\ 0, & \text{otherwise} \end{cases}$$

The probability density function $f_X(x,t)$ can be obtained as follows:

$$f_X(x,t) = \frac{f_\Theta(\theta_1)}{\left|\dfrac{dx(\theta_1)}{d\theta}\right|} + \frac{f_\Theta(\theta_2)}{\left|\dfrac{dx(\theta_2)}{d\theta}\right|}$$

because there are two values of $\theta \in [0, 2\pi]$ such that $x = A\cos(\omega_0 t + \theta)$, one value of θ is obtained where $0 \leq \omega_0 t + \theta \leq \pi$ and the other is obtained where $\pi \leq \omega_0 t + \theta \leq 2\pi$.

Now

$$\frac{dx}{d\theta} = -A\sin(\omega_0 t + \theta) = -A\sqrt{1 - \cos^2(\omega_0 t + \theta)}$$

$$= -\sqrt{A^2 - x^2}, \quad \text{for } 0 \leq \omega_0 t + \theta \leq \pi \quad \text{and} \quad |x| \leq A$$

and

$$\left|\frac{dx}{d\theta}\right|_{\theta=\theta_1} = \left|\frac{dx}{d\theta}\right|_{\theta=\theta_2}$$

43

$$\therefore f_X(x,t) = \frac{1}{2\pi \left|\frac{dx}{d\theta}\right|_{\theta=\theta_1}} + \frac{1}{2\pi \left|\frac{dx}{d\theta}\right|_{\theta=\theta_2}}$$

$$= \frac{1}{\pi \sqrt{A^2 - x^2}}, \quad \text{for } |x| < A$$

$$\therefore f_X(x,t) = \begin{cases} \dfrac{1}{\pi \sqrt{A^2 - x^2}}, & |x| < A \\ 0, & \text{otherwise} \end{cases}$$

(b) $m(t) = E[X(t)] = A \displaystyle\int_{-\infty}^{\infty} \cos(\omega_0 t + \theta) f(\theta) \, d\theta$

$$A \int_0^{2\pi} \cos(\omega_0 t + \theta) \frac{1}{2\pi} \, d\theta = 0$$

Alternatively,

$$m(t) = E[X(t)] = \int_{-\infty}^{\infty} x\, f(x,t)\, dx$$

$$= \int_{-A}^{A} x \frac{1}{\pi(A^2 - x^2)^{1/2}}\, dx = 0$$

(c) $E(X(t)^2) = \displaystyle\int_0^{2\pi} \overbrace{A^2 \cos^2(\omega_0 t + \theta)}^{x^2(\theta)} \overbrace{\frac{1}{2\pi}}^{f(\theta)} d\theta =$

$$\frac{A^2}{4\pi} \int_0^{2\pi} [1 + \cos 2(\omega_0 t + \theta)]\, d\theta = \frac{A^2}{4\pi}(2\pi) = \frac{A^2}{2}$$

$$\therefore \sigma_X^2 = E(X^2(t)) - E^2 X(t) = \frac{A^2}{2} - 0 = \frac{A^2}{2}$$

Remember that:

$$\sigma_X^2 = E(X^2) - E^2(X) = E(X^2)$$

$$= \int_{-\infty}^{\infty} x^2 f(x,t)\, dx = \int_{-A}^{A} x^2 \frac{1}{\pi \sqrt{1-x^2}}\, dx$$

$$= \frac{2}{\pi} \int_0^A \frac{x^2}{\sqrt{1-x^2}}\, dx = \frac{A^2}{2}$$

$$\left(\int \frac{x^2}{\sqrt{1-x^2}}\, dx = \left[-\frac{x}{2}\sqrt{1-x^2} + \frac{1}{2}\sin^{-1} x \right] \right)$$

2.4 SECOND AND HIGHER ORDER STATISTICS

For any arbitrary set of t-values $t_1, t_2, \ldots, t_n$ and random variables $X(t_1) = X_1, \ldots, X(t_n) = X_n$, we define the n-dimensional joint distribution as:

$$F(x_1, x_2, \ldots, x_n, t_1, t_2, \ldots, t_n) = P\{X_1 \leq x_1, \ldots, X_n \leq x_n\}$$

and the p.d.f. $f(x_1, \ldots, x_n, t_1, \ldots, t_n)$ is a function such that:

(1) $f(x_1, \ldots, x_n, t_1, \ldots, t_n) \geq 0$, for all $\mathbf{x} = (x_1, \ldots, x_n)^T$ and $t_1, \ldots, t_n$

(2) $F(x_1, \ldots, x_1, t_1, \ldots, t_n) =$

$$\int_{-\infty}^{x_1} \cdots \int^{x_n} f(x_1, \ldots, x_n, t_1, \ldots, t_n)\, dx_1 \ldots dx_n$$

Again, if F has a partial derivative with respect to $x_1, \ldots, x_n$, then:

$$f(x_1, \ldots, x_n, t_1, \ldots, t_n) = \frac{\partial^n F(x_1, \ldots, x_n, t_1, \ldots, t_n)}{\partial x_1\, \partial x_2 \ldots \partial x_n}$$

2.4.1 Autocorrelation; Covariance

The correlation between two waveforms from the same ensemble gives some useful information about the waveform. The first-order statistics do not yield all the information about the random process, since the first-order p.d.f. cannot indicate the dependence of the random process (signal) at two different times (remember that $X(t_1)$ and $X(t_2)$ are two different random variables). Thus, it would be advantageous to obtain a measure of relating the process $X(t_1)$ to $X(t_2)$.

For the real process $X(t)$, the autocorrelation function $R_X(t_1, t_2)$ is defined as:

$$R_X(t_1, t_2) = E\{X(t_1) X(t_2)\} = \int\int_{-\infty}^{\infty} x_1 x_2 f(x_1, x_2, t_1, t_2) \, dx_1 \, dx_2 \quad (2.5)$$

and it can easily be seen that it is a function of t_1 and t_2.

The corresponding covariance (autocovariance) of $X(t)$ is defined as:

$$C_X(t_1, t_2) = E\{[X(t_1) - m_1] [X(t_2) - m_2]\} \quad (2.6)$$

Note that:

$$C_X(t_1, t_2) = E\{X(t_1) X(t_2)\} - m_1 m_2 = R_X(t_1, t_2) - m_1 m_2$$

Thus, from (2.6), it is obvious that if $t_1 = t_2 = t$, then:

$$C_X(t, t) = \sigma^2_{X(t)}$$

More Definitions

If $X(t)$ and $Y(t)$ are two processes that (one or both) could be complex, then Eqs. (2.5) and (2.6) are generalized as follows:

$$R_X(t_1, t_2) = E\{[X(t_1) X^*(t_2)]\} \quad (2.7)$$

$$C_X(t_1, t_2) = E\{[X(t_1) - m_1] [X^*(t_2) - m_2^*]\}$$

$$= R_X(t_1, t_2) - m_1 m_2^* \tag{2.8}$$

where "*" denotes the complex conjugate.

The cross-correlation between $X(t)$ and $Y(t)$ is defined as:

$$R_{XY}(t_1, t_2) = E\{X(t_1) Y^*(t_2)\} \tag{2.9}$$

and its corresponding cross-covariance as:

$$C_{XY}(t_1, t_2) = E\{[X(t_1) - m_X] [Y^*(t_2) - m_Y^*]\}$$

$$= R_{XY}(t_1, t_2) - m_X m_Y^* \tag{2.10}$$

It is obvious that the nth order p.d.f. contains all the information about the first $(n-1)$ p.d.f. For example, we shall illustrate this point by the second-order p.d.f. Let $f(x_1, x_2, t_1, t_2)$ be given, then:

$$f(x_1, x_2, t_1, t_2) = f(x_1, t_1) f(x_2, t_2 \mid x_1, t_1)$$

We know that

$$f(x_1, t_1) = \int_{-\infty}^{\infty} f(x_1, x_2, t_1, t_2) \, dx_2$$

and the conditional p.d.f. can be obtained as the ratio of $f(x_1, x_2, t_1, t_2)$ over $f(x_1, t_1)$.

The correlation coefficient between $X(t_1)$ and $X(t_2)$ is defined as:

$$\rho_{12} = \frac{C_X(t_1, t_2)}{\sigma_{X_1} \sigma_{X_2}} \tag{2.11}$$

as expected.

2.5 STATIONARY PROCESSES

Definition 5

A stochastic process $X(t)$ is said to be strictly stationary if the entire family of its finite-dimensional distributions are invariant under a translation in t. That is, for given $t_1, t_2, \ldots, t_n$ time points, the distribution of $X(t_1 + \tau), X(t_2 + \tau), \ldots, X(t_n + \tau)$ (for $X(\cdot)$ real or complex) is independent of τ.

$$\therefore F(x_1, x_2, \ldots, x_n, t_1, \ldots, t_n) = F(x_1, x_2, \ldots, x_n, t_1 + \tau, \ldots, t_n + \tau)$$

(2.12)

for all n. Thus, we need to check Eq. (2.12) for all finite n. For $n = 1$, since $F(x, t) = F(x, t + \tau)$ or $f(x, t) = f(x, t + \tau)$ (if F is differentiable), then:

$$EX(t) = EX(t + \tau), \text{ for all } \tau \qquad (2.13)$$

which implies $EX(t)$ must be constant. For example, let $\tau = -t$, since $EX(t) = EX(t + \tau) = EX(t - t) = EX(0) =$ constant (that is, $EX(t) = EX(0)$ for all t as well).

Conclusion 1

For a strictly stationary process $EX(t)$ is constant and is independent of time t.

Now if $R_X(t_1, t_2)$ exists for all t_1 and t_2, then by definition of $R_X(t_1, t_2)$:

$$R_X(t_1, t_2) = E[X(t_1) X^*(t_2)] = E[X(t_1 + \tau) X^*(t_2 + \tau)] \qquad (2.14)$$

Equation (2.14) is true for any t_1, t_2 and τ. For the special case where $\tau = -t_1$, then $R_X(t_1, t_2)$ in Eq. (2.14) becomes:

$$R_X(t_1, t_2) = E[X(t_1) X^*(t_2)]$$

$$= E[X(t_1 + \tau) X^*(t_2 + \tau)]$$

$$= E[X(t_2 - t_1) X^*(t_2 - t_1)]$$

$$= E[X(0) X^*(t_2 - t_1)] \qquad (2.15)$$

Thus, we have shown that $R_X(t_1, t_2)$ is a function of time difference $t_2 - t_1$ (for the strictly stationary case).

Conclusion 2

It turns out that for the strictly stationary case we have $R_X(t_1, t_2)$ as a function of the time difference $t_2 - t_1$. From now on, when this condition prevails, we shall write $R_X(t_1, t_2)$ as $R(t_2 - t_1)$.

Conclusion 3

For strictly stationary processes, we have:

$$EX(t) = \text{constant} = m \qquad (2.16a)$$

$$EX(t_1) X^*(t_2) = R(t_2 - t_1) \qquad (2.16b)$$

The condition given by Eq. (2.16) is a consequence of a strictly stationary property (a necessary condition). In a strictly stationary process, we must have at our disposal all of the joint distribution functions for $k = 1, \ldots, n$ (finite n) and, in addition, they must satisfy:

$$F(x_1, \ldots, x_k, t_1, \ldots, t_k) = F(x_1, \ldots, x_k, t_1 + \tau, \ldots, t_k + \tau)$$

for all $k = 1, \ldots, n$ and all τ.

The above condition is very stringent. It turns out that very often the second-order statistics are sufficient to characterize many physical situations, which leads us to define some important terms.

Definition 6

The process $X(t)$ is stationary in the wide sense, if conditions (2.16a) and (2.16b) are satisfied.

2.5.1 Some Important Properties for the Wide-Sense Stationary Process $X(t)$

(1) $R(t_2 - t_1) = R^*(t_1 - t_2)$ or, equivalently, $R(t) = R^*(-t)$, since $R(t_2 - t_1) = E(X(t_1) X^*(t_2)) = E[(X(t_2) X^*(t_1)]^* = R^*(t_1 - t_2)$.

(2) Since $E|X(t)|^2 = E[X(t) X^*(t)] = R(0)$, then, $\sigma^2_{X(t)} = R(0) - m^2$, which is independent of time t.

(3) From the Cauchy-Schwarz inequality:

$$E[|X(t_1) X^*(t_2)|^2] \leq E[|X(t_1)|^2] E[|X(t_2)|^2] \Rightarrow |R(t)| \leq R(0),$$

for all t

Example 7

A quantized process has associated sample functions, where each sample function consists of sequences of pulses of unit width.

The pulse amplitudes take the binary numbers +1 and −1 with equal probability. *The successive amplitudes are independent.* Assume that the *starting point* of each sample function is random and uniformly distributed over a unit interval (denote the starting time as θ). Find the correlation function of $X(t)$.

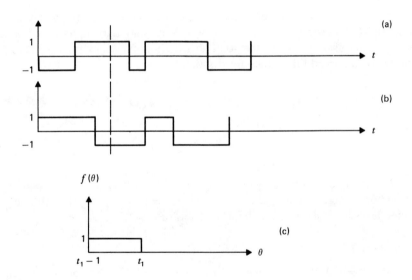

Solution

The random processes have discrete values of +1 and −1. Let $X(t_1) = i$ and $X(t_2) = j$, where i and j could be +1 or −1. Then,

$$R(t_1, t_2) = E[X(t_1) X(t_2)] = \sum_i \sum_j x_i x_2 P(i,j)$$

where

$$P(i,j) = P\{X(t_1) = i \text{ and } X(t_2) = j\}$$

$$\therefore R(t_1, t_2) = (1)(1) P(1,1) + (1)(-1) P(1,-1)$$

$$+ (-1)(-1) P(-1,-1) + (-1)(1) P(-1,1) \qquad (2.17)$$

Now if we obtain $P(i,j)$ for i and j corresponding to +1 or −1, we will be done. These probabilities are obtained as follows:

$$P(1,1) = P[X(t_2) = 1 \mid X(t_1) = 1] \, P[X(t_1) = 1]$$

For a sample function, let θ be the starting point of the pulse in which t_1 occurs (uniformly distributed, see part (c) of the above figure). Now t_2 either takes place during the same pulse as $t_1 < t_2$ (case 1) or during another pulse; we now write:

$$P[X(t_2) = 1 \mid X(t_1) = 1] = P[t_2 < \theta + 1] + \frac{1}{2} P[t_2 > \theta + 1]$$

(The 1/2 is used because outside the pulse, given $X(t_1) = 1$, it is equally likely that $X(t_2)$ be either +1 or −1.)

Now $P(1,1)$ can be written as:

$$P(1,1) = \left\{ P[t_2 < \theta + 1] + \frac{1}{2} P[t_2 > \theta + 1] \right\} \overbrace{P[X(t_1) = 1]}^{1/2}$$

$$= \frac{1}{2} \left\{ P[t_2 < \theta + 1] + \frac{1}{2} P[t_2 > \theta + 1] \right\}$$

$$= \begin{cases} \frac{1}{2}\left\{1 - (t_2 - t_1) + \frac{1}{2}(t_2 - t_1)\right\}, & \text{if } t_2 - t_1 \leq 1 \\ \frac{1}{2}\left[0 + \frac{1}{2}\right], & \text{if } t_2 - t_1 > 1 \end{cases}$$

Note that

$$P[t_2 < \theta + 1] = P[\theta > t_2 - 1] = 1 - P[\theta < t_2 - 1] = 1 - F(t_2 - 1)$$

and remember that $F(t) = t - (t_1 - 1) = t - t_1 + 1$

$$\therefore F(t_2 - 1) = t_2 - 1 - t_1 + 1 = t_2 - t_1 \text{ for the case } t_2 - t_1 \leq 1.$$

Because of symmetry, $P(1,1) = P(-1,-1)$. In a similar manner, we will find:

$$P(1,-1) = P(-1,1) = \begin{cases} \dfrac{1}{4}(t_2 - t_1), & \text{if } t_2 - t_1 \leq 1 \\ \dfrac{1}{4}, & \text{if } t_2 - t_1 > 1 \end{cases}$$

Now, for $\tau = t_2 - t_1$ (t_1 could be larger than t_2), the general case $R_X(\tau)$ can be found (see Eq. 2.17):

$$R_X(\tau) = \begin{cases} 1 - |\tau|, & \text{if } |\tau| \leq 1 \\ 0, & \text{if } |\tau| > 1 \end{cases}$$

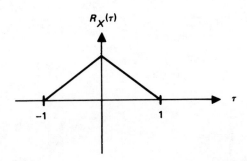

Henceforth, throughout the text, unless specified otherwise, by the stationarity of a process $X(t)$ we mean stationarity in the wide sense.

Definition 7

Two processes $X(t)$ and $Y(t)$ are uncorrelated if, given any t_1 and t_2, we have:

$$E[X(t_1) Y^*(t_2)] = m_X(t_1) m_Y^*(t_2) \qquad (2.18)$$

As a consequence of condition (2.18), we have:

$$C_{XY}(t_1, t_2) = E\{[X(t_1) - m_X(t_1)] [Y^*(t_2) - m_Y^*(t_2)]\}$$

$$= E[X(t_1) Y^*(t_2)] - m_X(t_1) m_Y^*(t_2)$$

$$= m_X(t_1) m_Y^*(t_2) - m_X(t_1) m_Y^*(t_2) = 0$$

Definition 8

If $E[X(t_1) Y^*(t_2)] = 0$, then we say $X(t)$ and $Y(t)$ are orthogonal.

Note that $C_{XY}(t_1, t_2) = 0$ implies that $[X(t_1) - m_X(t_1)]$ and $[Y(t_2) - m_Y(t_2)]$ are orthogonal processes.

2.6 CONTINUITY AND DIFFERENTIABILITY

The continuity of the process $X(t)$ with respect to t is restrictive. However, the continuity in the quadratic mean (mean square) is not as restrictive. We say the process $X(t)$ is continuous at $t = t_0$ in the quadratic mean (q.m.) if $E[|X(t_0)|^2]$ exists for $t = t_0$, and

$$\lim_{\epsilon \to 0} E\{|X(t_0) - X(t_0 + \epsilon)|^2\} = 0, \text{ for every } \epsilon \qquad (2.19)$$

If condition (2.19) holds for every $t \in [a, b]$, then we say $X(t)$ is continuous in the quadratic mean (mean square) in $[a, b]$. If condition (2.19) holds for $t \in (-\infty, \infty)$, we say $X(t)$ is continuous (in the q.m.) everywhere.

It is left as an exercise to verify the following claims.

Claim 1. $X(t)$ is continuous in the q.m. at $t = t_0$, if and only if the covariance $R(t_1, t_2)$ is continuous at every $t_1 = t_2 = t_0$ (diagonal point or element).

Note: In order to prove the above claim, we need to verify the important relationship:

$$E[|X(t+\epsilon) - X(t)|^2] = R(t+\epsilon, t+\epsilon) - R(t, t+\epsilon)$$

$$- R(t+\epsilon, t) + R(t, t) \quad (2.20)$$

The continuity in the q.m. is much weaker than the sample continuity. A classical counter example is the Poisson process:

$$P[X(t) = k] = \frac{(\lambda t)^k}{k!} \exp[-\lambda t]$$

where $X(t)$ is a staircase type and, therefore, discontinuous; however, $R(t_1, t_2) = \lambda \min(t_1, t_2)$, for all t_1 and t_2, is continuous, which implies $X(t)$ is continuous in the q.m. even though $X(t)$ is not continuous as a sample function.

If $X(t)$ satisfies:

$$\lim_{\epsilon \to 0} E\left[\left|\frac{X(t+\epsilon) - X(t)}{\epsilon} - X'(t)\right|^2\right] = 0 \quad (2.21)$$

We say $X'(t)$ is the derivative of $X(t)$ in the q.m. and we write:

$$\frac{X(t+\epsilon) - X(t)}{\epsilon} \xrightarrow[\epsilon \to 0]{\text{q.m.}} X'(t)$$

We can verify that (use Eq. 2.20):

$$E\left[\frac{X(t+\epsilon_1) - X(t)}{\epsilon_1} \frac{X^*(t+\epsilon_2) - X^*(t)}{\epsilon_2}\right]$$

$$= \frac{R(t+\epsilon_1, t+\epsilon_2) - R(t+\epsilon_1, t) - R(t, t+\epsilon_2) + R(t, t)}{\epsilon_1 \epsilon_2} \quad (2.22)$$

Claim 2. The derivative $X'(t)$ of $X(t)$ exists in the q.m. if and only if

$$\frac{\partial^2 R(t_1, t_2)}{\partial t_1 \, \partial t_2}$$

exists and is finite for $t_1 = t_2 = t$ (see Eq. 2.22) because, as ϵ_1 and $\epsilon_2 \to 0$, Eq. (2.22) becomes the second partial for $t_1 = t_2 = t$. Thus, the autocorrelation of $X'(t)$ is given by:

$$R_{X'X'}(t_1, t_2) = \frac{\partial^2 R_{XX}(t_1, t_2)}{\partial t_1 \, \partial t_2} \tag{2.23}$$

By direct calculation, it can also be shown that:

$$R_{XX'}(t_1, t_2) \triangleq E[X(t_1) X^{*'}(t_2)] = \frac{\partial R_{XX}(t_1, t_2)}{\partial t_2} \tag{2.24}$$

$$R_{X'X}(t_1, t_2) \triangleq E[X'(t_1) X^*(t_2)] = \frac{\partial R_{XX}(t_1, t_2)}{\partial t_1} \tag{2.25}$$

$$R_{X'X'}(t_1, t_2) \triangleq E[X'(t_1) X^{*'}(t_2)] = \frac{\partial R_{XX'}(t_1, t_2)}{\partial t_1} \tag{2.26}$$

If $X(t)$ is stationary, and utilizing $\tau = t_1 - t_2$, as well as Eqs. (2.24)–(2.26), we get:

$$R_{X'X'}(\tau) = -\frac{d^2 R_{XX}(\tau)}{d\tau^2} \tag{2.27}$$

From which:

$$R_{X'X'}(0) = E[|X'(0)|^2] = -\frac{d^2 R_{XX}(0)}{d\tau^2} \tag{2.28}$$

2.7 ERGODICITY AND STOCHASTIC INTEGRALS

In order to obtain the complete statistics of a process, the ensemble of sample functions is needed. Loosely speaking, a process is called ergodic if the complete statistics can be determined from any of the sample functions in the ensemble. Thus, a single member of the ensemble is assumed to represent the entire ensemble. Before giving a basic definition of ergodicity, the concept of stochastic integration is needed. Thus, we shall talk about the stochastic integrals.

2.8 STOCHASTIC INTEGRALS IN QUADRATIC MEAN

For the great majority of applications, we do not need the most general form of the stochastic integrals. Thus, we shall only consider two cases of integrals: Reimann integrals of the form:

$$A_1 = \int_a^b g(t) X(t) \, dt \qquad (2.29)$$

and Stieltjes integrals of the form:

$$A_2 = \int_a^b g(t) \, dX(t) \qquad (2.30)$$

where $[a, b]$ is the closed interval and is finite, $g(t)$ is a deterministic function, and $X(t)$ is a random process. For the sake of simplicity, assume $EX(t) = 0 = m(t)$. Thus,

$$R_X(t, u) = C_X(t, u)$$

Suppose $I = [a, b]$ is finite, and let the points $\alpha_1, \alpha_2, \ldots, \alpha_{m+1}$ define a partition, that is:

$$a = \alpha_1 < \alpha_2 \ldots < \alpha_{m+1} = b$$

Let S_1 and S_2 denote the sums corresponding to A_1 and A_2, respectively:

$$S_1 = \sum_{j=1}^{m} g(\alpha_j) X(\alpha_j)(\alpha_{j+1} - \alpha_j) \qquad (2.31)$$

$$S_2 = \sum_{j=1}^{m} g(\alpha_j)[X(\alpha_{j+1}) - X(\alpha_j)] \qquad (2.32)$$

Since S_1 and S_2 are summations of random variables, S_1 and S_2 are also random variables with $E(S_1) = E(S_2) = 0$ (because $EX(t) = 0$ by assumption for all t).

Now as $m \to \infty$ and the maximum of $(\alpha_{j+1} - \alpha_j) \to 0$, the limits of S_1 and S_2 exist (in the quadratic mean), that is,

$$A_1 \stackrel{q.m.}{=} \lim S_1 \qquad (2.33)$$

$$A_2 \stackrel{q.m.}{=} \lim S_2 \qquad (2.34)$$

where

$$m \to \infty \text{ and } \max (\alpha_{j+1} - \alpha_j) \to 0 \qquad (2.35)$$

Remark 6. From the above, we mean:

$$\lim E[|A_1 - S_1|^2] = 0$$

and

$$\lim E[|A_2 - S_2|^2] = 0$$

whenever condition (2.35) is satisfied.

Claim 3. It can be verified easily that if $R(t, u)$ is continuous over $[a, b] \times [a, b]$, and if $g(t)$ is such that the Reimann integral:

$$W_1 = \int_a^b \int_a^b g(t) \, g^*(u) \, R(t, u) \, dt \, du \qquad (2.36)$$

exists, then the integral A_1 exists in the quadratic mean (q.m.) and

$$E|A_1|^2 = W_1 \text{ and } E(A_1) = 0 \qquad (2.37)$$

Remember that $E(A_1) = 0$ (this was shown above).

Claim 4. Also, if $R(t, u)$ is of bounded variation ($|R(t, u)|$ has finite number of maximums and minimums over $[a, b] \times [a, b]$), and if $g(t)$ is such that the Stieltjes integral:

$$W_2 = \int_a^b \int_a^b g(t) \, g^*(t) \, dR(t, u) \qquad (2.38)$$

exists, then A_2 exists and

$$E|A_2|^2 = W_2 \text{ and } E(W_2) = 0 \qquad (2.39)$$

To prove (2.37) and (2.39), we consider another partition of $[a, b]$:

$$a = u_1 < u_2 \ldots < u_{m+1} = b$$

and we let S_1' and S_2' represent the sums corresponding to (2.31) and (2.32); then we can show (by utilizing the definitions) that:

$$E(S_1 S_1^{*\prime}) \to \int_a^b \int_a^b g(t) \, g^*(t) \, R(t, u) \, dt \, du \qquad (2.40)$$

where

$$m \to \infty \text{ and max } (\alpha_{j+1} - \alpha_j) \to 0 \text{ and max } (u_{j+1} - u_j) \to 0 \qquad (2.41)$$

Similarly,

$$E(S_2 S_2^{*\prime}) \to \int_a^b \int_a^b g(t) \, g^*(u) \, dR(t, u) \qquad (2.42)$$

as condition (2.40) is satisfied.

Remark 7. We have assumed that S_1 and S_2 converge in the q.m. It is easily shown that the limit in each case will be independent of the particular partition chosen.

Remark 8. If either S_1 or S_2 converges as $a \to -\infty$ and $b \to \infty$, then the limiting integrals are defined accordingly.

Remark 9. Since

$$A_1 = \int_a^b g(t) X(t) \, dt$$

then,

$$E|A_1|^2 = E[A_1 A_1^*] = E\left[\int_a^b g(t) X(t) \, dt \int_a^b g^*(u) X^*(u) \, du\right]$$

$$= E\left[\int_a^b \int_a^b g(t) g^*(u) X(t) X^*(u) \, dt \, du\right]$$

If we let the "expected value" E operate on the integrand, we would get the result given by (2.40). However, we can only do this if the appropriate conditions are satisfied.

Example 8

Let $g(t) = 1$ and $X(t)$ be a continuous real process on $[a, b]$; define:

$$q = \int_a^b X(t) \, dt$$

Find the mean and the variance of q. It is easy to show that the conditions of claim 3 are satisfied ($m(t)$ may not be zero, which was assumed for convenience in claim 3).

Solution

$$Eq = E\left[\int_a^b X(t)\,dt\right] = \int_a^b E(X(t))\,dt = \int_a^b m(t)\,dt \qquad (2.43)$$

Now we need to calculate $E(q^2)$, since $\sigma_q^2 = E(q^2) - E^2 q$:

$$q^2 = \int_a^b \int_a^b X(t)\,X(u)\,dt\,du$$

Again, the conditions of claim (3) are satisfied; thus,

$$E(q^2) = E\left[\int_a^b \int_a^b X(t)\,X(u)\,dt\,du\right]$$

$$= \int_a^b \int_a^b R(t,u)\,dt\,du$$

Thus, the variance becomes:

$$\sigma_q^2 = \int_a^b \int_a^b [R(t,u) - m(t)\,m(u)]\,dt\,du$$

$$= \int_a^b \int_a^b C(t,u)\,dt\,du \qquad (2.44)$$

Example 9

In Example 8, let

$$q = \frac{1}{2T}\int_{-T}^{T} X(t)\,dt$$

and assume $X(t)$ is stationary (wide sense); find σ_q^2.

Solution

From Eq. (2.43), we get:

$$Eq = \frac{1}{2T}\int_{-T}^{T} m\, dt = \frac{mt}{2T}\Big]_{-T}^{T} = m = \text{constant}$$

From Eq. (2.44), we get:

$$\sigma_q^2 = \frac{1}{4T^2}\int_{-T}^{T}\int_{-T}^{T} C(t-u)\, du\, dt \tag{2.45}$$

Equation (2.45) can be simplified much further.

Before proceeding with the simplification, let us review some simple mathematics (coordinate transformation). Let g_1 and g_2 be continuous (real) functions, such that:

$$x = g_1(w, z)$$

$$y = g_2(w, z)$$

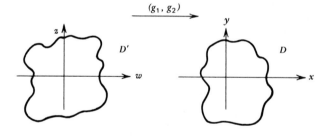

For example, (g_1, g_2) maps D' onto D. Then the following well known result is satisfied;

$$\iint_D f(x,y)\, dx\, dy = \iint_{D'} f(g_1(w,z), g_2(w,z)) \frac{\partial(x,y)}{\partial(w,z)}\, dw\, dz$$

$$\tag{2.46}$$

For any continuous real function $f(\cdot,\cdot)$, $\partial(x, y)/\partial(w, z)$ is the determinant of the Jacobian matrix:

$$\begin{bmatrix} \dfrac{\partial x}{\partial w} & \dfrac{\partial x}{\partial z} \\ \dfrac{\partial y}{\partial w} & \dfrac{\partial y}{\partial z} \end{bmatrix}$$

where the entries are continuous.

Application of the Above

Let $t_1 = t - u$ and $t_2 = t + u$. (This corresponds to a rotation of the axes by 45° and a scale change of $\sqrt{2}$.) The J (determinant) is determined:

$$\frac{\partial(t_1, t_2)}{\partial(t, u)} = \begin{vmatrix} 1 & -1 \\ 1 & 1 \end{vmatrix} = 2$$

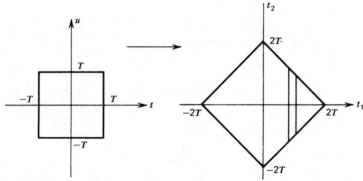

Thus,

$$J = \frac{1}{2} \text{ and}$$

$$\int_{-T}^{T}\int_{-T}^{T} C(t - u)\, dt\, du = \left(\frac{1}{2}\right) \int_{-2T}^{2T} \int_{-2T+|t_1|}^{2T-|t_1|} C(t_1)\, dt_1 dt_2$$

$$= \frac{1}{2} \int_{-2T}^{2T} dt_1\, C(t_1) \int_{-2T+|t_1|}^{2T-|t_1|} dt_2$$

$$= \frac{1}{2} \int_{-2T}^{2T} dt_1\, C(t_1)[2T - |t_1|]$$

$$= \int_{-2T}^{2T} (2T - |t_1|) C(t_1) dt_1$$

$$= \int_{-2T}^{2T} (2T - |\tau|) C(\tau) d\tau$$

where t_1 and τ are dummy variables.

Using this last result on Eq. (2.45) yields (dividing by $4T^2$):

$$\sigma_q^2 = \frac{1}{2T} \int_{-2T}^{2T} \left(1 - \frac{|\tau|}{2T}\right) C(\tau) d\tau \qquad (2.47)$$

Equation (2.47) is true for the complex $X(t)$ as well; however, for the real case, Eq. (2.47) further reduces to:

$$\sigma_q^2 = \frac{1}{T} \int_0^{2T} \left(1 - \frac{\tau}{2T}\right) C(\tau) d\tau \qquad (2.48)$$

2.9 DEFINITION OF ERGODICITY

Let $X(t)$ be a stationary process and assume that:

$$\lim_{T \to \infty} \frac{1}{2T} \int_{-T}^{T} x(t) dt$$

exists in the q.m. We say $X(t)$ is ergodic if:

$$\lim_{T \to \infty} \frac{1}{2T} \int_{-T}^{T} x(t) dt \stackrel{\text{q.m.}}{=} m \qquad (2.49)$$

That is,

$$E\left\{ \left| \frac{1}{2T} \int_{-T}^{T} x(t) dt - m \right|^2 \right\} \to 0, \text{ as } T \to \infty$$

From Example 9, we have:

$$Eq = E\left[\frac{1}{2T}\int_{-T}^{T} x(t)\, dt\right] = m$$

and utilizing Eq. (2.48) the variance of q is given by:

$$\sigma_q^2 = E\left\{\left|\frac{1}{2T}\int_{-T}^{T} x(t)\, dt - m\right|^2\right\} = \frac{1}{2T}\int_{-2T}^{2T}\left(1 - \frac{|\tau|}{2T}\right) C(\tau)\, d\tau$$

(2.50)

Thus, it is obvious that $X(t)$ is ergodic in the quadratic mean if and only if (see the above equation) the following is satisfied:

$$\frac{1}{2T}\int_{-2T}^{2T}\left(1 - \frac{|\tau|}{2T}\right) C(\tau)\, d\tau \to 0, \text{ as } T \to \infty \qquad (2.51)$$

2.10 SPECIAL PROCESSES WITH INDEPENDENT INCREMENTS

We shall discuss two important processes with independent increments. These processes, called Poisson and Wiener, are extremely important from both the theoretical and practical points of view. For example, the Poisson process serves as a very accurate model for many physical situations, such as radioactive disintegration, emission of electrons, occurrence of telephone calls, and various other events. The Wiener process is very important in the theoretical development of stochastic system modeling as with Kalman filtering, and is also used for a rigorous treatment of stochastic differential and integral equations. A rigorous approach for dealing with stochastic differential and integral equations is due to Ito. According to Ito's approach, the rules in ordinary calculus do not prevail in stochastic calculus. The Wiener process serves as a key element in the development of Ito's calculus. Because a full discussion of Ito calculus is beyond the scope of this book, see references [3], [6], [11], and [12] for further reading.

Definition 8

A stochastic process $X(t)$ with independent increments is a process whose increments over any two non-overlapping intervals are independent random variables. That is, if we let the points t_a, t_b, t_c, and t_d be specified such that

$$t_a < t_b \leqslant t_c < t_d$$

then

$$P[X(t_b) - X(t_a) = \gamma \mid X(t_d) - X(t_c) = \delta] = P[X(t_b) - X(t_a) = \gamma] \quad (2.52)$$

Note that if we define $Y = X(t_b) - X(t_a)$ and $Z = X(t_d) - X(t_c)$, then Y and Z would be independent random variables.

Example 10

Let $X(t)$ be a stochastic process with independent increments given via the figure below. The random variables defined by $X(t_2) - X(t_1)$ and $X(t_4) - X(t_3)$

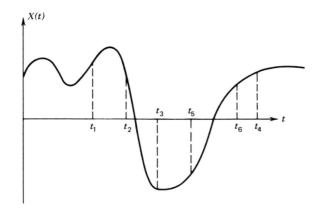

are independent. However, the random variables defined by $X(t_4) - X(t_3)$ and $X(t_6) - X(t_5)$ are not independent, because the intervals (t_3, t_4) and (t_5, t_6) have a nonempty intersection.

2.10.1 Poisson Process

Let $X(t)$ be a process with independent increments, then $X(t)$ is called a Poisson process if the following assumptions are made:

(1) $X(0) = 0$, $X(t)$ is a nondecreasing step function, which increases with a jump 1 at each discontinuity.

(2) The probability that exactly n jumps (points) occur on the interval (t_1, t_2) is given by:

$$P[X(t_2) - X(t_1) = n] = \frac{\exp[-\lambda(t_2 - t_1)]\lambda^n}{n!} \quad (2.53)$$

when λ is a positive constant. The probability that exactly n points occur in the time interval $(0, t)$ depends on both n and t, but not the position of the interval $(0, t)$. For example, in Eq. (2.53) if we set $t = t_2 - t_1$, then

$$P[X(t) = n] = \frac{\exp(-\lambda t)\lambda^n}{n!} \quad (2.54)$$

Also see the figure below.

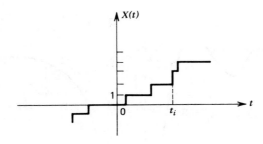

It is left as an exercise to verify that (see problems 2.11 and 2.12):

$$EX(t) = \lambda t$$

$$R(t_1, t_2) = \begin{cases} \lambda t_1 + \lambda^2 t_1 t_2, & \text{if } t_2 \geqslant t_1 \\ \lambda t_2 + \lambda^2 t_1 t_2, & \text{if } t_2 \leqslant t_1 \end{cases} \quad (2.55)$$

Thus, it is obvious that the Poisson process $X(t)$ is nonstationary. We can also calculate the variance σ_X^2 from Eq. (2.55):

$$\sigma_X^2 = E[X^2(t)] - [EX(t)]^2 = R(t, t) - (\lambda t)^2$$

$$= \lambda t + \lambda^2 t^2 - \lambda^2 t^2 = \lambda t$$

Hence, for the Poisson process we have:

$$\sigma_X = \sqrt{\lambda t} = \sqrt{m_X} \qquad (2.56)$$

2.10.2 Wiener or Brownian Motion Process

Another important process is the Wiener or Brownian motion process. Let $X(t)$ be a process with independent increments. Then, $X(t)$ is a Wiener process if the following conditions are satisfied:

(1) $X(t)$ is a normal process for every t

(2) $EX(t) = 0$ and $R(t_1, t_2) = \min\{t_1, t_2\}$

Equation (1.41) represents the general p.d.f. form of a normal vector, where we showed that the only information needed to define the density function is the mean and the covariance of the vector. Applying (1.41) to our specific case, the Wiener process, one can easily show that if $t_1 < t_2 \cdots < t_n$, then the p.d.f. of the vector

$$\mathbf{X} = (X_1, X_2, \ldots, X_n)^T$$

where T denotes the transpose, is given by:

$$f_X(x_1, x_2, \ldots, x_n, t_1, t_2, \ldots, t_n) = \prod_{i=1}^{n} \frac{\exp\left[\frac{-1}{2} \frac{(x_i - x_{i-1})^2}{t_i - t_{i-1}}\right]}{\sqrt{2\pi(t_i - t_{i-1})}}$$

$$(2.57)$$

Also see problems 2.13 and 2.14.

From the above equation we can derive some interesting properties. For example,

$$E[(X(t_i) - X(t_{i-1}))^2] = (t_i - t_{i-1})$$

or, in general

$$E[(X(t_b) - X(t_a))^2] = |t_b - t_a| \qquad (2.58)$$

for any t_b and t_a.

Consequently, for $t \geq 0$, $EX^2(t) = t$.

2.10.3 Definition of White Noise and the Levy Process

Let $Y(t) = \sigma X(t)$, then it is very simple to show (see problem 2.14) that:

$$EY(t) = 0$$

and

$$E[(Y(t_2) - Y(t_1))^2] = \sigma^2 |t_2 - t_1|$$

We know from Eq. (2.23) that:

$$R_{X'}(t_1, t_2) = \frac{\partial^2 R_X(t_1, t_2)}{\partial t_1 \partial t_2}$$

If we let $t_1 = t_2 = t$, then there would exist a discontinuity at t, where for $t_1 = t_2$, the variance is σ^2, and is zero otherwise. Thus,

$$E\left[\frac{dY(t_2)}{dt_2} \frac{dY(t_1)}{dt_1}\right] = \sigma^2 \delta(t_2 - t_1) \tag{2.59}$$

where $\delta(t_2 - t_1)$ is the delta function. The relation given by Eq. (2.59) is not very rigorous; however, a rigorous treatment may be obtained in references [6] and [13]. Appendix A gives an accurate treatment of the delta functions.

Definition 9

Let a process $N(t)$ be defined as:

$$N(t) = \frac{dY(t)}{dt}$$

We say $N(t)$ is a white noise process if Eq. (2.59) is satisfied.

Definition 10

Another process which has independent increments and is used in proving some theorems in stochastic calculus is the Levy process. The Levy process $X(t)$ satisfies the following conditions:

(1) The discontinuities, say at t_0 are of the kind that

$$\lim_{\epsilon \to 0} X(t_0 + \epsilon) \neq \lim_{\epsilon \to 0} X(t_0 - \epsilon)$$

(2) $P[\,|X(t_2) - X(t_1)| > 0\,] = 0$ as $t_2 \to t_1$.

The Levy process was briefly discussed because this process has independent increments and will be used in stochastic calculus. Since we shall not use this process in the subsequent chapters, it is not needed to elaborate on it further.

EXERCISES

2.1 Sketch a few samples of the process $X(t)$ given by:

$$X(t) = A \sin(\omega t + \Theta)$$

(a) If A is a random variable uniformly distributed over $[-1,1]$.
(b) If ω is random and uniformly distributed over $[0,\pi]$.
(c) If Θ is random and uniformly distributed over $(0,2\pi)$.

2.2 Obtain the mean and the variance of each process in problem 2.1.

2.3 Let the sample function process $X(t)$ be given by:

$$x(t) = a \cos(\omega_0 t + \theta)$$

Assume a is deterministic and θ is a value of the random variable Θ, where Θ is uniformly distributed over $[0,\pi/2]$. Find the mean, variance, and the autocorrelation function of $X(t)$.

2.4 Let the sample functions of a process $X(t)$ be given by:

$$x(t) = \cos(\omega_0 t + \theta)$$

where θ is uniformly distributed over $[0,2\pi]$. Obtain the p.d.f. of the process, and comment on the stationarity of the process (in the wide sense).

2.5 Let $Z(t) = X(t) Y(t)$ be real processes. Assume that $X(t)$ and $Y(t)$ are independent stationary processes (wide sense); then:

(a) Obtain $R_Z(\tau) = R_X(\tau) R_Y(\tau)$.

(b) If the processes $P(t) = X(t) - m_X$ and $Q(t) = Y(t) - m_Y$ with the corresponding

$$R_P(\tau) = \exp(-a|\tau|)$$

and

$$R_Q(\tau) = \exp(-b|\tau|)$$

where a and b are both positive, then obtain $R_Z(\tau)$.

2.6 Let $X(t)$ be a wide-sense stationary random process with no periodic components. Assume $X(t)$ and $X(t + \tau)$ are uncorrelated as $|\tau|$ becomes large. Show:

$$R_X(\tau) = m_X^2$$

2.7 If $X(t)$ and $Y(t)$ are independent random wide-sense stationary processes and $Z(t)$ and $W(t)$ are such that:

$$Z(t) = X(t) + Y(t); \quad W(t) = 2X(t) + Y(t)$$

Then find $R_Z(\tau)$, $R_W(\tau)$, $R_{ZW}(\tau)$, and $R_{WZ}(\tau)$.

2.8 Consider the process $X(t) = l(t)Y$, where $l(t)$ is a deterministic complex function (non-random), and Y is a random variable. Assume that we have a constraint on $X(t)$ such that $X(t)$ is of mean zero and is wide-sense stationary. Then perform the following:

(a) Determine the restriction on $l(t)$.

(b) Obtain the most general form of $l(t)$ that satisfies the requirement.

2.9 A process $Y(t)$ satisfies:

$$\dot{Y} + Y = X(t), \quad t > 0$$

where $Y(0) = 2$, $m_X = 1$, and $R_X = 1 + \exp(-|\tau|)$. Find the following:

(a) m_Y.

(b) $R_{XY}(t_1, t_2)$, for t_1 and $t_2 > 0$.

(c) $R_{YY}(t_1, t_2)$, for t_1 and $t_2 > 0$.

(d) Comment on the stationarity of R_{YY}.

2.10 Assume $C_X(\tau)$ of the process $X(t)$ satisfies:

$$\int_{-\infty}^{\infty} |C_X(\tau)| \, d\tau < \infty$$

Show that

$$\lim_{T \to \infty} \frac{1}{2T} \int_{-T}^{T} R_X(\tau) \, d\tau = m_X^2$$

2.11 Given the processes $X(t)$ and $N(t)$ such that

$$X(t) = b + N(t)$$

where b is a constant, $E(N) = 0$, and N is stationary, show that if $\hat{b}$ is given via

$$\hat{b} = \frac{1}{T} \int_0^T x(t) \, dt$$

it will satisfy

$$E(\hat{b}) = b$$

and

$$\text{variance of } \hat{b} = \frac{1}{T} \int_{-T}^{T} \left(1 - \frac{|\tau|}{T}\right) R_N(\tau) \, d\tau$$

2.12 Given $t_b > t_a$ and $Y = X(t_b) - X(t_a)$, we know that $P[Y = n] = \exp[-\lambda(t_b - t_a)] \lambda^n / n!$, if $X(t)$ is a Poisson process. Show that

$$EY = \lambda(t_b - t_a)$$

and

$$EY^2 = \lambda^2(t_b - t_a) + \lambda(t_b - t_a)$$

2.13 As a consequence of the above problem, show that

$$EX(t) = \lambda t$$

and

$$R(t_2, t_1) = \begin{cases} \lambda t_1 + \lambda^2 t_1 t_2, & \text{if } t_2 \geq t_1 \\ \lambda t^2 + \lambda^2 t_1 t_2, & \text{if } t_2 \leq t_1 \end{cases}$$

2.14 If $X(t)$ is a Wiener process, verify that for $t_2 \geq t_1$:

$$P[X(t_2) - X(t_1) < \lambda] = \frac{1}{\sqrt{2\pi(t_2 - t_1)}} \int_{-\infty}^{\infty} \exp\left[\frac{-\theta^2}{2(t_2 - t_1)}\right] d\theta$$

2.15 In the above problem, if we define $Y(t) = \sigma X(t)$, we say $Y(t)$ is a Wiener process that is denormalized. Show that:

$$EY(t) = 0$$

and

$$E[(Y(t_2) - Y(t_1))^2] = \sigma^2(t_2 - t_1)$$

CHAPTER 3
POWER SPECTRUM OF STATIONARY PROCESSES

Before discussing the power spectrum, which is defined for the wide-sense stationary, we need to familiarize ourselves with some basic concepts and definitions.

3.1 CLASSIFICATION OF SYSTEMS

Heuristically speaking, a system refers to a modeling of a physical phenomenon (which is idealized in some sense). We shall visualize a system via a black box which has many inputs and many outputs (vector input-output).

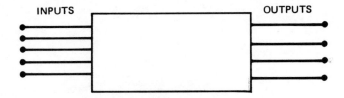

The input-output is often indicated symbolically by:

$$Y(t) = L\, U(t) \qquad (3.1)$$

where $U(t)$ is a vector-valued input, $Y(t)$ is a vector-valued output, and L is an "operator" relating the input to the output. The operator L depends on the particular physical model.

The notation given in Eq. (3.1) requires some clarification. If the input is defined on the time interval $[t_1, t_2]$, then we shall denote the corresponding input as $U_{[t_1,t_2]}$. To make Eq. (3.1) meaningful, we must assume that before an input is applied to the system, the output must be zero and thereafter uniquely determined by the input. For example, if the input $U_{[t_1,t_2]}$ is applied to the system, where both the input and the output are assumed to be zero for $t < t_1$, then such a system is called relaxed at t_1. In applied problems, it is often assumed that a system is relaxed at $t = -\infty$. In this chapter we shall also assume every system is relaxed at $t = -\infty$, and simply call such systems relaxed. Under this assumption the characterization of systems by Eq. (3.1) is well defined. The concept of linearity can now be defined.

Definition 1

We say the system is linear if the operator L is linear, i.e., the following conditions are satisfied:

$$L(\alpha U) = \alpha L(U) \qquad (3.2)$$

where α is any scalar, and

$$L(U_1 + U_2) = L(U_1) + L(U_2) \qquad (3.3)$$

for any inputs U_1 and U_2. Equivalently, Eqs. (3.2) and (3.3) can be combined into one equation:

$$L(\alpha U_1 + \beta U_2) = \alpha L(U_1) + \beta L(U_2) \qquad (3.4)$$

for any pair of scalars α and β.

In the following examples assume the inputs and the outputs are one-dimensional.

Example 1

Consider

$$y(t) = \dot{u}(t) = \frac{d}{dt} u(t)$$

We know $L = d/dt$ and the conditions of linearity are satisfied.

Example 2

$y(t) = u^2(t)$ does not correspond to a linear system since:

$$L[\alpha u_1(t) + \beta u_2(t)] = [\alpha u_1(t) + \beta u_2(t)]^2$$

$$\neq \alpha L(u_1(t)) + \beta L(u_2(t)) = \alpha u_1^2(t) + \beta u_2^2(t)$$

Example 3

Consider the electric circuit given below.

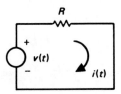

Let $v(t)$ be the input and $i(t)$ be the output. Then, the output is given by:

$$i(t) = \frac{1}{R} v(t) \qquad (3.5)$$

It is easy to verify that the system is linear.

Example 4

In the previous example change R to an inductor L and assume $i(-\infty) = 0$. Then,

$$i(t) = \frac{1}{L} \int_{-\infty}^{t} v(\lambda) \, d\lambda \qquad (3.6)$$

and the system is also linear (left as an exercise).

Definition 2

A system is called instantaneous if its output at any given time t is at most a function of the input at the same time.

Definition 3

A system is called dynamic if it is not instantaneous. Example 3 is instantaneous and Example 4 is dynamic.

Definition 4

A system, whose output at time t is completely determined from the input in the closed interval $[t - T, t]$, where $T \geq 0$, is said to have a memory T. Thus, if $T \neq 0$, the system is dynamic, otherwise it is instantaneous. In Example 4, the memory is infinite.

Definition 5

A system is realizable or causal if its output $y(t)$ does not depend on the future value of the input. Thus, $y(t)$ can be determined from the past (and the present) information of $u(\lambda)$ (i.e., $\lambda \leq t$ and not on $\lambda > t$).

Definition 6

A dynamic system is said to be lumped if it can be characterized by a set of differential equations for the continuous case (and difference equations for the discrete case).

In the classical characterization of a linear system, any lumped linear system (assume scalar inputs and outputs) can be represented by:

$$y(t) = \int_{-\infty}^{\infty} h(t, \tau) u(\tau) \, d\tau \qquad (3.7)$$

where $h(t, \tau) = L\,\delta(t - \tau)$ = response to a unit impulse function applied at time τ.

If the linear system is causal, then:

$$h(t, \tau) = 0, \quad \text{for} \quad \tau > t \qquad (3.8)$$

Otherwise, $y(t)$ would depend on $u(\tau)$ for $\tau > t$ (future value). Thus, it would not be realizable. Hence, Eq. (3.7) for causal systems can be written as:

$$y(t) = \int_{-\infty}^{t} \mathbf{h}(t, \tau) \, u(\tau) \, d\tau \qquad (3.9)$$

Definition 7

A system is time-invariant if the time translation of the input causes the same time translation in the output. That is, if

$$y(t) = L \, u(t)$$

then $u(t - \lambda)$ would correspond to $y(t - \lambda)$.

It is easy to verify in a linear time-invariant system that the impulse response $\mathbf{h}(t, \tau) = L(\delta(t - \tau))$ becomes:

$$\mathbf{h}(t, \tau) = L \, \delta(t - \tau) = h(t - \tau) \qquad (3.10)$$

where $\mathbf{h}$ and h are two different functions.

Thus, the linear time-invariant system is entirely specified by a response to a single unit impulse, which can be applied at any given time t. For the sake of simplicity, we shall assume the time $t = 0$. Hence,

$$h(t) = L \, u(t) \qquad (3.11)$$

For a time-invariant linear system given by Eq. (3.7), one can write:

$$y(t) = \int_{-\infty}^{\infty} h(t - \tau) \, u(\tau) \, d\tau \qquad (3.12)$$

Equation (3.12) is of a well known form, called the "convolution integral," and it is denoted in the literature by $h * u$. We are going to talk more about $h * u$ in later sections.

Remark 1. Since the integral given by Eq. (3.12) is the limit of a summation (definition of Reimann integral), we can think of the output $y(t)$ (signal) to be resolved into unit impulses. For example, consider a finite interval $[-T, T]$ and finite unit of pulses (steps) with width $\Delta\tau$ occurring at $t = k\Delta\tau$, for $k = 0, \pm1, \pm2, \ldots, \pm N = T/\Delta\tau$ (see sketch).

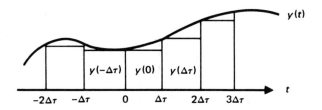

The summation

$$\sum_{k=-N}^{N} y(k\Delta\tau) P_{\Delta\tau}(t - k\Delta\tau) \Delta\tau$$

where $P_{\Delta\tau}(t - k\Delta\tau)$ is a unit pulse with width $\Delta\tau$. The height of the unit pulse is $1/\Delta\tau$ to make the pulse area equal to one. As $\Delta\tau \to 0$, $N \to \infty$, and $T \to \infty$; then, if the limit of the above summation exists, it must be equal to $y(t)$ given via Eq. (3.12), i.e.,

$$\int_{-\infty}^{\infty} h(t - \tau) u(\tau) d\tau$$

Discussion

Physical systems are characterized by models consisting of idealized elements. Choosing an appropriate model which characterizes all features of the physical system is very important and also very difficult. In general, a model of the physical system may be expressed mathematically via integro-differential equations and is generally nonlinear. The complete treatment of nonlinear systems is extremely difficult; therefore, we try to do the next best thing: approximate the nonlinear system with a linear system.

The classical method of describing a linear system is by the impulse response method. Even though the solution of the linear model is known, its treatment in the time domain for the time-varying case is not simple. If the linear model is time-invariant, we can use a transformation (such as Laplace or Fourier) to convert the complicated integro-differential equations into simple algebraic equations (frequency domain). It is of extreme importance to emphasize that the transforms can be used to great advantage only in the time-invariant linear systems. In the nonlinear and time-varying cases the transforms cannot be utilized to advantage.

It is very easy to imagine a situation where we transmit a random process $X(t)$ (signal) through a linear or a nonlinear system. However, if $X(t)$ is transmitted through a time-invariant linear system, we shall use Fourier transforms to simplify the calculations. The Fourier transform is also used for the decomposition of signal power, which will be defined in the following sections.

3.2 FREQUENCY SPECTRA AND FOURIER TRANSFORMS

Before developing the concept of the power spectrum of a stationary process, let us give some intuitive discussion of Fourier transforms and series. If the reader is not familiar with these concepts, he is advised to review Appendices C and D. In this section, however, a relatively non-rigorous approach is adopted for intuitive appeal only.

Let us start by asking ourselves the following question: Is there an input signal which will pass through a time-invariant system without changing shape? The answer is "yes" and is an exponential function exp (λt), where λ is, in general, a complex constant. If we choose a special form of exp (λt), namely, exp $(j\omega t)$, then the output $y(t)$ would be proportional to the input, i.e., $y(t) = H(j\omega)$ exp $(j\omega t)$, where $H(j\omega)$ is the so-called "system function." Since the characterization of the exponential functions of the general form exp (λt) (or exp $(j\omega t)$) is very simple, it is desired to resolve any general function $f(t)$ in terms of the exponentials whenever possible. Obviously, one such case is the representation of a periodic signal $f(t)$ in terms of exp $(j\omega t)$ (Fourier series).

A periodic signal $f(t)$ (not yet a random process) with a period T under a set of conditions (Dirichlet, see Appendix C) may be resolved into a series of complex functions over $[-T/2, T/2]$. The resolution is given by:

$$f(t) = \sum_{n=-\infty}^{\infty} C_n \exp(jn\omega_0 t) \qquad (3.13)$$

where $\omega_0 = 2\pi/T$, $t \in [-T/2, T/2]$, and the values of C_n are given by:

$$C_n = \frac{1}{T} \int_{-T/2}^{T/2} f(t) \exp(-jn\omega_0 t)\, dt \qquad (3.14)$$

Recall in Eq. (3.14) that C_n is, in general, complex and can be written as:

$$C_n = |C_n| \exp(j\theta_n) \tag{3.15}$$

where C_n and θ_n are functions of $\omega = n\omega_0$.

The essential information about the harmonics in a periodic signal consists of the magnitudes, phase angles, and frequencies. It is easy to see that all the information about $f(t)$ is incorporated in C_n and $\omega_0 = 2\pi/T$, since once these quantities are known, so is $f(t)$. The real amplitudes $|C_n|$ and the phases θ_n can be represented graphically as a function of $\omega = n\omega_0$, $n = 0, \pm 1, \pm 2, \ldots$. The collection of the graphs is called the frequency spectra (discrete). Typical amplitude and phase spectra are shown in Fig. 1. It is easily verified that $|C_n|$ is an even function of ω, and θ_n is an odd function of ω (left as an exercise). The reader may verify for himself that, for real signals $f(t)$,

$$C_n = C_{-n}^*$$

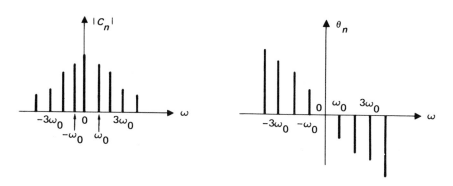

Fig. 3-1. Typical Phase and Amplitude Spectra

3.2.1 The Fourier Transform

Now suppose that the function $f(t)$ is defined over the infinite interval $(-\infty, \infty)$ and that it is no longer periodic. Then it is still possible, under certain conditions, to resolve the nonperiodic function into complex exponential functions of the form $\exp(j\omega t)$. The intuitive argument is to reduce the spacing ω_0 between the components of a periodic signal. Denote the spacing by $\Delta\omega = \omega_0 = 2\pi/T$ (radians per second). We shall continue to consider $|C_n|$

as a discrete function of $n\omega_0$. Since (see Eq. 3.14) $|C_n| \to 0$ as $T \to \infty$, we shall define a new variable $G(jn\omega_0) = G(jn\Delta\omega)$:

$$G(jn\Delta\omega) \triangleq \frac{C_n}{\Delta\omega/2\pi} = TC_n$$

As $T \to \infty$ and $\Delta\omega \to 0$, $n\Delta\omega$ approaches a continuous variable ω and:

$$G(\omega) = \int_{-\infty}^{\infty} f(t)\exp(-j\omega t)\,dt \qquad (3.16)$$

and $f(t)$ can be written as:

$$f(t) = \lim_{\Delta\omega \to 0} \sum_{n=-\infty}^{\infty} C_n \exp(j\omega_0 t) = \lim_{\Delta\omega \to 0} \sum_{n=-\infty}^{\infty} \frac{C_n}{\frac{\Delta\omega}{2\pi}} \exp(jn\Delta\omega t)\,\Delta\omega$$

As $\Delta\omega \to 0$, $n\Delta\omega$ approaches a continuous variable ω, such that

$$f(t) = \frac{1}{2\pi}\int_{-\infty}^{\infty} G(\omega)\exp(j\omega t)\,d\omega \qquad (3.17)$$

Equations (3.16) and (3.17) are called the Fourier transform pair. Equation (3.16) is, in general, a complex function of ω. As an exercise the reader can show that for real functions $f(t)$:

$$F^*(\omega) = F(-\omega) \qquad (3.18)$$

Also, the reader will find it instructive to verify the transform pairs given in the appendix on Fourier transforms.

If we use $f = \omega/2\pi$, and let $P(f) = G(2\pi f)$, then

$$P(f) = G(2\pi f) = \int_{-\infty}^{\infty} f(t)\exp(-j2\pi ft)\,dt \qquad (3.19)$$

and

$$f(t) = \int_{-\infty}^{\infty} P(f) \exp(j2\pi ft) \, df \qquad (3.20)$$

Thus,

$$\int_{-\infty}^{\infty} P(f) \exp(j2\pi ft) \, df = \frac{1}{2\pi} \int_{-\infty}^{\infty} G(\omega) \exp(j\omega t) \, d\omega \qquad (3.21)$$

Equations (3.19) and (3.20) are also called the Fourier transform pair.

3.3 POWER SPECTRA

We know that if $G(\omega)$ corresponding to the nonperiodic function $f(t)$ exists, then we can verify (see Appendix C) that:

$$\int_{-\infty}^{\infty} |f(t)|^2 \, dt = \frac{1}{2\pi} \int_{-\infty}^{\infty} |G(\omega)|^2 \, d\omega \qquad (3.22)$$

holds (Parseval's relation for Fourier transform).

Let $f(t)$, for example, represent the voltage across a resistance of 1 ohm. Then the instantaneous power $p(t)$ defined by $p(t) = v(t) i(t)$, where $v(t)$ is the voltage and $i(t)$ is the current through the resistance. Thus, the dissipated energy in the resistance (which is the integral of $p(t)$) is given by:

$$\int_{-\infty}^{\infty} |v(t)|^2 \, dt = \int_{-\infty}^{\infty} |f(t)|^2 \, dt = \int_{-\infty}^{\infty} |G(\omega)|^2 \frac{d\omega}{2\pi} \qquad (3.23)$$

The average power P_{AV} is defined by:

$$P_{AV} = \lim_{T \to \infty} \frac{1}{2T} \int_{-T}^{T} |f(t)|^2 \, dt \qquad (3.24)$$

It is possible that total energy be infinite and the average power to be finite. Note that $|G(\omega)|^2$ from Eq. (3.23) represents the density spectrum, except for the constant $1/2\pi$.

Now let us consider $X(t)$ to be a real stationary random process. Define $X_T(t)$ such that

$$X_T(t) = \begin{cases} X(t), & |t| \leq T \\ 0, & |t| > T \end{cases} \quad (3.25)$$

and let its Fourier be denoted by $\chi_T(\omega)$, i.e.,

$$\chi_T(\omega) = \int_{-\infty}^{\infty} X_T(t) \exp(-j\omega t)\, dt = \int_{-T}^{T} X(t) \exp(-j\omega t)\, dt \quad (3.26)$$

We can see that, as $T \to \infty$, the signal $X_T(t) \to X(t)$. Utilizing Eq. (3.24), the average power of $X(t)$ for $t \in [-T, T]$ is given by:

$$\frac{1}{2T} \int_{-T}^{T} [X(t)]^2\, dt = \int_{-\infty}^{\infty} \frac{|\chi_T(\omega)|^2}{2T} \frac{d\omega}{2\pi}$$

where from Eq. (3.23), $|\chi_T(\omega)|^2/(2T)$ represents the *power spectral density*. However, the power spectrum $S(\omega)$ of $X(t)$ is defined as:

$$S(\omega) = \lim_{T \to \infty} \frac{1}{2T} E[|\chi_T(\omega)|^2] \quad (3.27)$$

Now $S(\omega)$, by utilizing Eqs. (3.26) and (3.27), becomes:

$$S(\omega) = \lim_{T \to \infty} \frac{1}{2T} E[\chi_T(\omega)\, \chi_T^*(\omega)]$$

$$= \lim_{T \to \infty} \frac{1}{2T} E\left\{ \left[\int_{-T}^{T} X(t) \exp(-j\omega t)\, dt\right] \left[\int_{-T}^{T} X(t) \exp(j\omega t)\, dt\right] \right\}$$

The above equation can also be written as:

$$S(\omega) = \lim_{T \to \infty} \frac{1}{2T} \left[\int_{-T}^{T} \int_{-T}^{T} R_X(t - u) \exp(-j(t - u)) \, dt \, du \right]$$

where, from Example 9 of Chapter 2, we get:

$$S(\omega) = \lim_{T \to \infty} \int_{-2T}^{2T} R_X(\tau) \left[1 - \frac{|\tau|}{2T} \right] \exp(-j\omega\tau) \, d\tau$$

$$= \int_{-\infty}^{\infty} R_X(\tau) \exp(-j\omega\tau) \, d\tau \tag{3.28}$$

Thus, for a stationary process, $S(\omega)$ is the Fourier transform of $R_X(\tau)$:

$$R_X(\tau) = \frac{1}{2\pi} \int_{-\infty}^{\infty} S(\omega) \exp(j\omega\tau) \, d\omega \tag{3.29}$$

For a real process $X(t)$, $R_X(\tau) = R_X(-\tau)$, Eq. (3.29) becomes:

$$R_X(\tau) = \frac{1}{2\pi} \int_{-\infty}^{\infty} S(\omega) [\cos \omega\tau + j \sin \omega\tau] \, d\omega$$

$$= \frac{1}{2\pi} \int_{-\infty}^{\infty} S(\omega) \cos \omega\tau \, d\omega$$

$$= \frac{1}{\pi} \int_{0}^{\infty} S(\omega) \cos \omega\tau \, d\omega \tag{3.30}$$

Definition 8

The power spectrum of any stationary random process $X(t)$ (real or complex) is denoted by $S(\omega)$ and is given by:

$$S(\omega) = \int_{-\infty}^{\infty} R(\tau) \exp(-j\omega\tau)\, d\tau$$

where $R(\tau)$ is related to $S(\omega)$ by Eq. (3.29) for the general complex case, where Eq. (3.30) corresponds to the real case.

3.3.1 Examples

Before getting involved with the examples, a method of calculation for the bilateral Laplace transform is discussed. Assume the bilateral Laplace transform $F_B(s)$ of $f(t)$ exists in some region, say, for $\sigma_1 < \text{Re } s < \sigma_2$. Then,

$$F_B(s) = \int_{-\infty}^{\infty} f(t) \exp(-st)\, dt$$

$$= \int_{-\infty}^{0} f(t) \exp(-st)\, dt + \int_{0}^{\infty} f(t) \exp(-st)\, dt$$

$$= \int_{0}^{\infty} f(-t) \exp[-(-s)t]\, dt + \int_{0}^{\infty} f(t) \exp(-st)\, dt$$

$$= \mathscr{L}[f(-t)]_{(\text{replace } s \text{ by } -s)} + \mathscr{L}[f(t)]$$

where $\mathscr{L}$ is the one-sided Laplace transform.

Example 5

Find $F_B(s)$ of $f(t) = \dfrac{1}{2} \exp(-|t|)$.

Solution

For $t > 0$,

$$f(t) = \frac{1}{2}\exp(-t)$$

Hence,

$$\mathcal{L}[f(t)] = F(s) = \frac{1/2}{s+1}, \text{ for Re } s > -1$$

Now, for $t < 0$,

$$f(t) = \frac{1}{2}\exp(t)$$

which implies that:

$$\mathcal{L}[f(-t)] = \mathcal{L}\left[\frac{1}{2}\exp(-t)\right]\Big|_{\text{(replace } s \text{ by } -s)} = \frac{1/2}{s+1}\Big|_{\text{(replace } s \text{ by } -s)}$$

$$= \frac{1/2}{-s+1}, \text{ for Re } s < 1.$$

$$\therefore F_B(s) = \mathcal{L}[f(-t)]\Big|_{\text{(replace } s \text{ by } -s)} + \mathcal{L}[f(t)]$$

$$= \frac{1/2}{-s+1} + \frac{1/2}{s+1} = \frac{1}{1-s^2}$$

and the region of convergence is $-1 < \text{Re } s < 1$.

Remark 2. The Fourier transform $\mathcal{F}(\omega)$ of $f(t)$ is obtained by replacing s by $j\omega$. Hence, $\mathcal{F}(\omega) = 1/(1 + \omega^2)$.

Example 6

Given the stochastic differential equation:

$$\dot{x} = -x(t) + u(t)$$

where $x(0) = 0$ and $E[u(t)u(\tau)] = \delta(t - \tau)$, the solution of $x(t)$ is given by:

$$x(t) = x(0)\exp(-t) + \int_0^t \exp[-(t - \lambda)]\, u(\lambda)\, d\lambda$$

$$= \int_0^t \exp[-(t - \lambda)]\, u(\lambda)\, d\lambda$$

and

$$E[x(t)x(t + \tau)] = E\left\{\int_0^t \exp[-(t - \lambda)]\, u(\lambda)\, d\lambda \int_0^{t+\tau} \exp[-(t + \tau - \xi)]\, u(\xi)\, d\xi\right\}$$

$$= \int_0^t \int_0^{t+\tau} \exp[-(2t + \tau - \xi - \lambda)]\, E[u(\lambda)\, u(\xi)]\, d\lambda\, d\xi$$

$$= \int_0^t \int_0^{t+\tau} \exp[-(2t + \tau - \xi - \lambda)]\, \delta(\xi - \lambda)\, d\lambda\, d\xi$$

$$= \int_0^t \exp[-(2t + \tau - 2\xi)]\, d\xi = \frac{1}{2}\exp[-(2t + \tau)]\exp(2\xi)\Big]_0^t$$

$$= \frac{1}{2}\exp(-\tau) - \frac{1}{2}\exp[-(2t + \tau)], \text{ if } \tau \geq 0$$

Now, as $t \to \infty$,

$$R_X(\tau) = E[X(t)\, X(t + \tau)] = \frac{1}{2}\exp(-\tau), \text{ for all } \tau \geq 0$$

Similarly, we can find $R_X(\tau)$ for $\tau \leq 0$:

$$R_X(\tau) = \frac{1}{2} \exp(\tau)$$

$$R_X(\tau) = \frac{1}{2} \exp(-|\tau|), \text{ for all } \tau$$

To obtain $S_X(\omega)$, we can either use Example 6 or the direct definition of the Fourier transform. Thus,

$$S_X(\omega) = \frac{1}{2} \int_{-\infty}^{\infty} \exp(-|\tau|) \exp(-j\omega\tau) \, d\tau$$

$$= \frac{1}{2} \left[\frac{2}{\omega^2 + 1} \right] = \frac{1}{\omega^2 + 1}$$

Example 7

Suppose $S_X(\omega)$ of a process $X(t)$ is given by:

$$S_X(\omega) = \frac{1}{\omega^2 + 1}$$

Find $R_X(\tau)$ by the Theory of Residues.

Before completing this example, let us give an informal discussion of the inversion formula.

Let $f(t)$ be a given function such that its Fourier transform $\mathcal{F}(\omega)$ exists. Then, for a fixed positive $\sigma > 0$, the Fourier transform of $\exp(-\sigma t) f(t)$ also exists and is given by:

$$\int_{-\infty}^{\infty} f(t) \exp(-\sigma t) \exp(-j\omega t) \, dt = \int_{-\infty}^{\infty} f(t) \exp[-(\sigma + j\omega) t] \, dt$$

Denote the integral as $F(\sigma + j\omega)$. Thus, $f(t)\exp(-\sigma t)$ is given by:

$$f(t)\exp(-\sigma t) = \mathscr{F}^{-1}[F(\sigma + j\omega)]$$

$$= \frac{1}{2\pi}\int_{-\infty}^{\infty} F(\sigma + j\omega)\exp(j\omega t)\,d\omega$$

Multiplying both sides by $\exp(\sigma t)$ (σ is constant), we get:

$$f(t) = \frac{1}{2\pi}\int_{-\infty}^{\infty} F(\sigma + j\omega)\exp[(\sigma + j\omega)t]\,d\omega$$

Making the change of variable $s = \sigma + j\omega$, we obtain:

$$f(t) = \frac{1}{2\pi j}\int_{\sigma-j\infty}^{\sigma+j\infty} F_B(s)\exp(st)\,ds$$

$$= \begin{cases} \sum \text{residues of } F_B(s)\exp(st) \text{ at singularities to left of line chosen,} & \text{for } \tau \geqslant 0 \\ -\sum \text{residues of } F_B(s)\exp(st) \text{ at singularities to right of line chosen,} & \text{for } \tau < 0 \end{cases}$$

The equivalent bilateral transform corresponding to $S_X(\omega)$ is denoted by $S_B(s)$ and is obtained from $S_X(\omega)$ by substituting $\omega = s/j$.

Now applying the inversion formula to Example 7:

$$S_B(s) = S(s/j) = \frac{1}{1-s^2} = \frac{-1}{s^2-1} = \frac{-1}{(s-1)(s+1)}$$

where $S_B(s)$ exists for $-1 < \text{Re } s < 1$. Now,

$$R(\tau) = \frac{1}{2\pi j}\int_{c-j\infty}^{c+j\infty} S_B(s)\exp(s\tau)\,ds, \text{ where } -1 < c < 1$$

$$= \begin{cases} \sum \text{residues of } S_B(s)\exp(s\tau) \text{ at poles of } S_B(s), \quad \text{for } \tau \geq 0 \\[1em] -\sum \text{residues of } S_B(s)\exp(s\tau) \text{ at poles of } S_B(s), \quad \text{for } \tau < 0 \end{cases}$$

$$= \begin{cases} \left.\dfrac{-1(1+s)\exp(s\tau)}{(s-1)(1+s)}\right|_{s=-1} = \dfrac{-1}{-2}\exp(-\tau) \\[1em] \qquad\qquad\qquad\qquad = \dfrac{\exp(-\tau)}{2}, \text{ for } \tau \geq 0 \\[1em] -\left.\dfrac{-(s-1)\exp(s\tau)}{(s-1)(1+s)}\right|_{s=1} = \left.\dfrac{\exp(s\tau)}{1+s}\right|_{s=1} \\[1em] \qquad\qquad\qquad\qquad = \dfrac{\exp(\tau)}{2}, \text{ for } \tau < 0 \end{cases}$$

$$= \frac{1}{2}\exp(-|\tau|) \; \forall \; \tau$$

Example 8

If $S(\omega)$ is a power spectrum of a given process, show that $d^2S/d\omega^2$ is not a power spectrum.

Solution

$$S(\omega) = \int_{-\infty}^{\infty} R(\tau)\exp(-j\omega\tau)\,d\tau$$

which implies:

$$\frac{d^2 S(\omega)}{d\omega^2} = \int_{-\infty}^{\infty} [-\tau^2 R(\tau)] \exp(-j\omega\tau)\, d\tau \triangleq \mathcal{F}\{-\tau^2 R(\tau)\}$$

Now, if $d^2 S(\omega)/d\omega^2$ is a power spectrum, we must have $[-\tau^2 R(\tau)]$ as an autocorrelation function. Let $G(\tau) = -\tau^2 R(\tau)$. If $G(\tau)$ is an autocorrelation, then we would always have:

$$|G(\tau)| \leq G(0), \text{ for all } \tau$$

However, $G(0) = 0$ and

$$0 = G(\tau) \leq G(0)$$

cannot be always satisfied.

Example 9

$X(t) = \cos(\omega_0 t + \theta)$, $\theta \in [0, 2\pi]$, is uniformly distributed. Find $S_X(\omega)$.

Solution

From Example 6, Chapter 2:

$$R(\tau) = EX(t)X(t+\tau) = \frac{1}{2}\cos \omega_0 \tau$$

$$\therefore S_X(\omega) = \mathcal{F}\left\{\frac{1}{2}\cos \omega_0 \tau\right\} = \frac{\pi}{2}[\delta(\omega - \omega_0) + \delta(\omega + \omega_0)]$$

$$= \frac{\pi}{2(2\pi)}[\delta(f - f_0) + \delta(f + f_0)]$$

$$= \frac{1}{4}[\delta(f - f_0) + \delta(f + f_0)]; \quad 2\pi f_0 = \omega_0$$

Example 10

In Example 8 of Chapter 2, the autocorrelation function $R(\tau)$ was given by:

$$\begin{cases} 1 - |\tau|, & |\tau| \leq 1 \\ 0, & |\tau| > 1 \end{cases}$$

Find $S_X(\omega)$.

Solution

$$S_X(\omega) = \int_{-\infty}^{\infty} (1 - |\tau|) \exp(-j\omega\tau) \, d\tau = \int_{-1}^{1} (1 - |\tau|) \exp(-j\omega\tau) \, d\tau$$

$$= 2 \int_{0}^{1} (1 - \tau) \cos \omega\tau \, d\tau$$

$$= \frac{2}{\omega^2} (1 - \cos \omega) = \frac{\sin^2\left(\frac{\omega}{2}\right)}{\left(\frac{\omega}{2}\right)^2}$$

$$= \left(\frac{\sin \frac{\omega}{2}}{\frac{\omega}{2}}\right)^2$$

3.4 MAJOR RESULT

In what follows, we shall show that a function $R(\tau)$ which has a Fourier transform $S(\omega)$ is an autocorrelation function of a stationary random process $X(t)$ if and only if $S(\omega) \geq 0$ for all ω, where $X(\cdot)$ is continuous in the quadratic mean (q.m.). In order to prove this major result, we need to prove some important results given by Theorems 1 and 2, which will appear in the sequel. We shall assume $X(\cdot)$ is continuous in the q.m. unless specified otherwise.

Theorem 1 (Bochner's Theorem)

The function $R(\tau)$ is an autocorrelation function of a stationary process $X(t)$ if and only if $R(\tau)$ is nonnegative definite.

We have already shown that if $R(\tau)$ is an autocorrelation function, then it is nonnegative definite, since for any collections of $t_1, t_2, \ldots, t_n$ (time) and complex parameters $\alpha_1, \alpha_2, \ldots, \alpha_n$:

$$\sum_{i,k=1}^{n} R(t_j - t_k) \alpha_j \alpha_k^* = E \left| \sum_{j=1}^{n} X(t_j) \alpha_j \right|^2 \geq 0 \qquad (3.31)$$

However, the converse is more complicated and will not be proven here. (For the proof, see Gnedenko, *Theory of Probability*, Chelsea publication, 1962.)

Theorem 2

A function $R(\tau)$ with the corresponding $S(\omega)$ is nonnegative definite (autocorrelation) if and only if it can be represented by:

$$R(\tau) = \frac{1}{2\pi} \int_{-\infty}^{\infty} \exp(j\omega\tau) S(\omega) \, d\omega$$

where $S(\omega)$ is never negative (i.e., $S(\omega) \geq 0$, for all ω).

The proof is relatively complicated and will be eliminated here; for a proof see the same reference shown in Theorem 1.

As a special case of the Fourier transform pair $R(\tau)$ and $S(\omega)$, we have:

$$S(0) = \int_{-\infty}^{\infty} R(\tau) \, d\tau \qquad (3.32)$$

and

$$R(0) = \frac{1}{2\pi} \int_{-\infty}^{\infty} S(\omega) \, d\omega = \int_{-\infty}^{\infty} S(\omega) \frac{d\omega}{2\pi} \qquad (3.33)$$

and $R(0)$ is the average power by definition, i.e.,

$$R(0) = E[|X(t)|^2]$$

Definition 9

A stationary process $X(t)$ whose power spectrum $S(\omega)$ is constant for all ω is called a white-noise process. If $S(\omega) = W_0$ = constant, we obtain:

$$R(\tau) = \frac{1}{2\pi} \int_{-\infty}^{\infty} W_0 \exp(j\omega\tau) \, d\omega = W_0 \, \delta(\tau) \qquad (3.34)$$

Hence $R(0)$, which is the average power, becomes infinite at $\tau = 0$. Thus, we conclude that the white noise process is a mathematical function that is very useful in practical applications. For example, it is convenient to utilize white noise as an approximation to an actual process whose power spectrum is flat (constant) over a frequency band.

In application problems such as those that occur in control and communication, we are faced with physical noise sources which are added to the signal as a lump sum. The power spectrum of the overall noise is essentially flat up to frequencies much higher than those that are significant for the signal and the system.

3.5 INPUT-OUTPUT RELATIONS

Very often we confront a situation where we pass a stationary process $X(t)$ through a time-invariant system, and are interested in determining the output (along with its statistics).

Consider the (bounded) sample function $X(t)$ from the ensemble $\{X(t)\}$ which is applied to a time-invariant system with impulse response $h(t)$ (see sketch) and the output $Y(t)$.

We know $Y(t)$ can be written as:

$$Y(t) = \int_{-\infty}^{\infty} h(\lambda) \, X(t - \lambda) \, d\lambda \qquad (3.35)$$

Now let us find $R_Y(\tau)$.

From Eq. (3.35), we have:

$$Y(t + \tau) = \int_{-\infty}^{\infty} h(u)\, X(t + \tau - u)\, du$$

Thus, $R(\tau) = E[Y(t)\, Y(t + \tau)]$ can be written as:

$$R(\tau) = E\left[\int_{-\infty}^{\infty} h(\lambda)\, X(t - \lambda)\, d\lambda \int_{-\infty}^{\infty} h(u)\, X(t + \tau - u)\, du \right]$$

(3.36)

Rewriting Eq. (3.36) and taking the expectation inside yields:

$$R_Y(\tau) = E[y(t)\, y(t + \tau)] = \int_{-\infty}^{\infty}\int_{-\infty}^{\infty} h(\lambda)\, h(u)\, E[X(t - \lambda)\, X(t + \tau - u)]\, d\lambda\, du$$

$$= \int_{-\infty}^{\infty}\int_{-\infty}^{\infty} h(\lambda)\, h(u)\, R_X(\tau + \lambda - u)\, d\lambda\, du$$

$$= h(-\tau) * h(\tau) * R_X(\tau) \qquad (3.37)$$

Now if $S_Y(\omega)$, $H(\omega)$, and $S_X(\omega)$ exist, we can apply the Fourier transform to Eq. (3.37) to get:

$$S_Y(\omega) = \mathscr{F}\{h(-\tau)\} \cdot \mathscr{F}\{h(t)\} \cdot \mathscr{F}\{R_X(\tau)\}$$

$$= H^*(\omega)\, H(\omega)\, S_X(\omega) = |H(\omega)|^2\, S_X(\omega) \qquad (3.38)$$

which is an important relationship yielding $S_Y(\omega)$ in terms of $S_X(\omega)$ and the system transfer function $H(\omega)$.

Remark 3. From Eq. (3.37) it is obvious that $E[y(t)\, y(t + \tau)]$ is a function of τ alone, and also due to stationarity of $X(t)$, $EX(t) = m = $ constant, which implies $EY(t)$ is also constant (see Eq. 3.35). Hence, $Y(t)$ is stationary (wide sense).

Remark 4.

$$R_Y(0) = E[|Y(t)|^2] = \frac{1}{2\pi} \int_{-\infty}^{\infty} S_Y(\omega)\, d\omega$$

$$= \frac{1}{2\pi} \int_{-\infty}^{\infty} |H(j\omega)|^2\, S_X(\omega)\, d\omega$$

(3.39)

Remark 5. The results are also true for the complex stochastic processes.

3.6 INPUT-OUTPUT OF MULTIPLE TERMINALS

Suppose we have two time-invariant systems characterized by their impulse responses $h_1(\cdot)$ and $h_2(\cdot)$, respectively (see sketch):

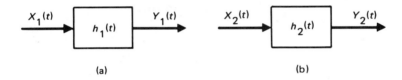

(a)   (b)

where $X_1(t)$ and $X_2(t)$ are sample functions from $\{X(t)\}$, which as before is assumed to be a stationary ensemble.

Let us calculate $R_{Y_1 Y_2}(\tau)$. As before $Y_1(t)$ and $Y_2(t)$ can be written as:

$$Y_1(t) = \int_{-\infty}^{\infty} h_1(\lambda)\, X_1(t - \lambda)\, d\lambda \qquad (3.40)$$

$$Y_2(t) = \int_{-\infty}^{\infty} h_2(u)\, X_2(t - u)\, du \qquad (3.41)$$

and a simple calculation (similar to the previous case) of $R_{Y_1 Y_2}(\tau)$ would lead to:

$$R_{Y_1 Y_2}(\tau) = E[Y_1(t) Y_2(t + \tau)]$$

$$= E\left[\int_{-\infty}^{\infty} h_1(\lambda) X_1(t - \lambda) d\lambda \int_{-\infty}^{\infty} h_2(u) X_2(t + \tau - u) du\right]$$

$$= \int_{-\infty}^{\infty} \int_{-\infty}^{\infty} h_1(\lambda) h_2(u) R_{X_1 X_2}(\tau + \lambda - u) d\lambda \, du \quad (3.42)$$

where $R_{X_1 X_2}(\tau)$ is the cross-correlation of $X_1(t)$ and $X_2(t)$. Hence, once again:

$$R_{Y_1 Y_2}(\tau) = h_1(-\tau) * h_2(\tau) * R_{X_1 X_2}(\tau) \quad (3.43)$$

Thus, assuming that the appropriate Fourier transforms exist, we obtain:

$$S_{Y_1 Y_2}(\omega) = H_1(-\omega) H_2(\omega) S_{X_1 X_2}(\omega)$$

$$= H_1^*(\omega) H_2(\omega) S_{X_1 X_2}(\omega) \quad (3.44)$$

which is a very general result, relating the input spectrum of $R_{X_1 X_2}(\tau)$ to the output spectrum $S_{Y_1 Y_2}(\omega)$.

Note that as a special case of Eq. (3.44), if we let $X_1 = X_2$ and $h_1 = h_2$ (which implies $Y_1 = Y_2$), we obtain Eq. (3.38). Note that Eq. (3.44) is also true for complex processes.

Remark 6. The reader may verify for himself that if $X_1(t)$ and $X_2(t)$ are uncorrelated, so are $Y_1(t)$ and $Y_2(t)$.

Discussion

In applications, $H_1(\omega)$ and $H_2(\omega)$ very often have finite bandwidths, i.e., $H_1(\omega) = 0$ for some ω_0 such that $|\omega| > \omega_0$ and, similarly, $H_2(\omega) = 0$ for some ω_1 such that $|\omega| > \omega_1$. It is obvious that if $H_1(\omega)$ and $H_2(\omega)$ have nonoverlapping spectra, then

$$H_1(-\omega) H_2(\omega) = 0$$

which would yield:

$$S_{Y_1 Y_2}(\omega) = 0$$

or

$$R_{Y_1 Y_2}(\tau) = 0$$

In that case, the processes $Y_1(t)$ and $Y_2(t)$ would be orthogonal.

A very important consequence of the above is that if $X(t)$ is transmitted through an ideal filter, i.e.,

$$|H(\omega)| = \begin{cases} A_0, & \text{for } |\omega| < \omega_0 \\ 0, & \text{otherwise} \end{cases}$$

then the output signal $Y(t)$ and the *signal suppressed by the filter* would be orthogonal. That is, if $X(t)$ has a frequency content beyond ω_0, it is going to be suppressed by $H(\omega)$, and the suppressed portion is orthogonal to $Y(t)$.

Example 11

A white-noise voltage source $X(t)$ with power spectrum $S_X(\omega) = K_0$ is applied to an RLC network (see sketch). Assuming that the system (circuit) is at rest at $t = 0$ (no transients), determine $S_Y(\omega)$.

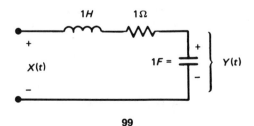

Solution

$$H(\omega) = \frac{\frac{1}{j\omega C}}{R + j\omega L + \frac{1}{j\omega C}} = \frac{\frac{1}{j\omega}}{1 + j\omega + \frac{1}{j\omega}}$$

We know that:

$$S_Y(\omega) = |H(\omega)|^2 S_X(\omega) = |H(\omega)|^2 K_0$$

$H(\omega)$ can be calculated from the above as follows:

$$|H(\omega)|^2 = \left| \frac{\frac{1}{j\omega}}{1 + j\omega - \frac{j}{\omega}} \right|^2 = \frac{\frac{1}{\omega^2}}{1 + \left(\omega - \frac{1}{\omega}\right)^2}$$

$$\therefore S_Y(\omega) = \frac{\frac{1}{\omega^2}}{1 + \left(\omega - \frac{1}{\omega}\right)^2} K_0 = \frac{K_0}{\omega^4 - \omega^2 + 1}$$

3.7 SAMPLING THEOREM

The sampling theorem (due to C. E. Shannon*) is very important and has produced some unexpected results. The utilization of this theorem is prevalent in control and communication theory. It must be emphasized that the sampling theorem, whether we are dealing with deterministic or stochastic signals, will only hold for *band-limited signals,* that is, signals whose Fourier transforms are identically zero beyond a finite band of frequencies. In order to develop this concept, we shall first deal with a signal $X(t)$, which is deterministic. To be more precise, we shall state the theorem.

Theorem 3

Given a deterministic signal $X(t)$ whose Fourier transform $\mathcal{X}(\omega)$ is zero beyond $|\omega| > \omega_c$ rad/s (see sketch):

$$\mathcal{X}(\omega) = 0, \quad \text{for all } |\omega| > \omega_c$$

---

*C. E. Shannon, "Communication in Presence of Noise," *Proc. IRE,* Jan. 1947.

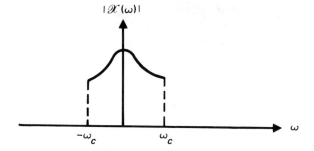

Then $X(t)$ can be completely and uniquely recovered by its values sampled at uniform intervals of $T = \pi/\omega_c$ seconds (or smaller), and it is given by:

$$X(t) = \sum_{n=-\infty}^{\infty} X(nT) \frac{\sin[\omega_c(t - nT)]}{\omega_c(t - nT)} \qquad (3.45)$$

Proof

There are several ways of proving this important theorem, but we shall give the simplest proof.

From the inverse Fourier transform, we obtain:

$$X(t) = \frac{1}{2\pi} \int_{-\infty}^{\infty} \mathscr{X}(\omega) \exp(j\omega t)\, d\omega = \frac{1}{2\pi} \int_{-\omega_c}^{\omega_c} \mathscr{X}(\omega) \exp(j\omega t)\, d\omega \qquad (3.46)$$

Now, assume that $\mathscr{X}(\omega)$ is a part of a periodic function $\mathscr{X}^+(\omega)$ (see sketch), such that:

$$\mathscr{X}(\omega) = \mathscr{X}^+(\omega), \quad \text{if } |\omega| < \omega_c$$

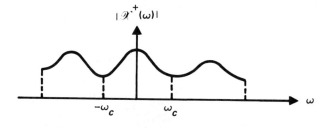

Hence, for $|\omega| < \omega_c$ (see Appendix D).

$$\mathscr{X}(\omega) = \sum_{n=-\infty}^{\infty} b_n \exp(jn\omega T) \qquad (3.47)$$

where

$$T = \frac{2\pi}{2\omega_c} = \frac{\pi}{\omega_c}$$

and b_n is given by:

$$b_n = \frac{T}{2\pi} \int_{-\omega_c}^{\omega_c} \mathscr{X}(\omega) \exp(-jn\omega T) \, d\omega \qquad (3.48)$$

If we substitute $t = -nT$ in Eq. (3.46), we obtain:

$$X(-nT) = \frac{1}{2\pi} \int_{-\omega_c}^{\omega_c} \mathscr{X}(\omega) \exp(-jn\omega T) \, d\omega$$

$$= \frac{1}{T}\left(\frac{T}{2\pi}\right) \int_{-\omega_c}^{\omega_c} \mathscr{X}(\omega) \exp(-jn\omega T) \, d\omega$$

Now, utilizing the definition of b_n from Eq. (3.48), we get from the above equation:

$$X(-nT) = \frac{1}{T} b_n$$

or, equivalently,

$$b_n = T X(-nT)$$

Using the above in Eq. (3.47) yields:

$$\mathcal{X}(\omega) = T \sum_{n=-\infty}^{\infty} X(nT) \exp(-jn\omega T) \qquad (3.49)$$

Now, if we substitute Eq. (3.49) into Eq. (3.46), we obtain:

$$X(t) = \sum_{n=-\infty}^{\infty} X(nT) \frac{T}{2\pi} \int_{-\omega_c}^{\omega_c} \exp[j\omega(t - nT)] \, d\omega$$

$$= \sum_{n=-\infty}^{\infty} X(nT) \frac{\sin[\omega_c(t - nT)]}{\omega_c(t - nT)} \qquad (3.50)$$

which is exactly the result we are after.

Remark 7. If we substitute $T = \pi/\omega_c$, then

$$\frac{\sin[\omega_c(t - nT)]}{\omega_c(t - nT)} = \frac{\sin\left[\omega_c\left(t - \frac{n\pi}{\omega_c}\right)\right]}{\omega_c\left(t - \frac{n\pi}{\omega_c}\right)} = \frac{\sin(\omega_c t - n\pi)}{(\omega_c t - n\pi)} \qquad (3.51)$$

Remark 8. The function

$$\frac{\sin[\omega_c(t - nT)]}{\omega_c(t - nT)} = \frac{\sin(\omega_c t - n\pi)}{(\omega_c t - n\pi)}$$

is an "interpolation function" which is multiplied by $X(nT)$ and is summed over all n to yield $X(t)$.

Now we shall discuss the case where $X(t)$ is a stochastic process. We will show that the result given by Eq. (3.51) holds for the stochastic case in the quadratic mean (q.m.). That is,

$$X(t) \stackrel{\text{q.m.}}{=} \sum_{n=-\infty}^{\infty} X(nT) \frac{\sin(\omega_c t - n\pi)}{(\omega_c t - n\pi)}$$

or, equivalently,

$$E\left[\left|X(t) - \sum_{n=-\infty}^{\infty} X(nT)\frac{\sin(\omega_c t - n\pi)}{(\omega_c t - n\pi)}\right|^2\right] = 0 \qquad (3.52)$$

Before completing the proof, we shall discuss some properties concerning the periodicity of $X(t)$ and $R_X(\tau)$. In what follows, $X(t)$ is assumed to be wide-sense stationary unless specified otherwise.

Discussion

In Appendix D we discuss the periodicity of the deterministic signals and construct an infinite-dimensional vector space L_2 with its subspace H that was spanned by the set

$$\{\exp(jn\omega_0 t)\}_{n=-\infty}^{\infty}$$

For stochastic signals we shall modify Appendix D. If we change the norm

$$\|f\|^2 = (f,f) = \int_{-T/2}^{T/2} |f(t)|^2 \, dt$$

in the appendix for the deterministic case to:

$$\|X\|^2 = (X,X) = E[|X(t)|^2]$$

for the stochastic case, all of the results will hold. Thus, the norm for the stochastic case is the quadratic mean or the mean square. Now if a stochastic process $X(t)$ is periodic (almost everywhere and not in the quadratic mean as yet), i.e.,

$$X(t) = X(t+T) = X(t+nT)$$

Then it is very easy to verify that $R_X(\tau)$ is also periodic, since

$$R_X(\tau) = E[X(t+\tau)X^*(t)] = E[X(t+\tau+nT)X^*(t)] = R_X(\tau+nT)$$

$$(3.53)$$

Hence, the periodicity of $X(t)$ implies the periodicity of $R_X(\tau)$. However, if $R_X(\tau)$ is periodic with the period T, i.e.,

$$R_X(\tau + T) = R_X(\tau)$$

then (left as an exercise), it can be shown that $X(t)$ is periodic in the quadratic mean and can be expanded into a Fourier series:

$$X(t) \stackrel{q.m.}{=} \sum_{n=-\infty}^{\infty} a_n \exp(jn\omega_0 t), \quad \omega_0 = \frac{2\pi}{T} \qquad (3.54)$$

where a_n's are the usual Fourier coefficients and are pairwise orthogonal, i.e.,

$$E[a_n a_m^*] = 0, \text{ for all } n \neq m$$

We can also write $R(\tau)$ by a Fourier series given by:

$$R_X(\tau) = \sum_{n=-\infty}^{\infty} C_n \exp(jn\omega_0 t), \quad \omega_0 = \frac{2\pi}{T} \qquad (3.55)$$

where the C_n's are again the Fourier coefficients, and the a_i's and the C_i's are related via:

$$C_n = E[|a_n|^2]$$

Now, if we use the Fourier transform on $R_X(\tau)$, we get:

$$S_X(\omega) = 2\pi \sum_{n=-\infty}^{\infty} C_n \delta(\omega - n\omega_0) \qquad (3.56)$$

3.7.1 Application of Sampling Theorem to Autocorrelation Functions

Let as assume that the autocorrelation function $R_X(\tau)$ has a band-limited spectrum $S_X(\omega)$. Application of Eq. (3.45) would give rise to:

$$R_X(\tau) = \sum_{n=-\infty}^{\infty} R_X(nT) \frac{\sin(\omega_c \tau - n\pi)}{(\omega_c \tau - n\pi)} \qquad (3.57)$$

It would be easy to verify that if $X(t)$ is a band-limited stochastic process, then

$$X(t) \stackrel{q.m.}{=} \sum_{n=-\infty}^{\infty} X(nT) \frac{\sin(\omega_c t - n\pi)}{(\omega_c t - n\pi)} \qquad (3.58)$$

The following proof is from reference [1]. To prove Eq. (3.58), we show:

$$E\left[\left(X(t) - \sum_{n=-\infty}^{\infty} X(nT) \frac{\sin(\omega_c t - n\pi)}{(\omega_c t - n\pi)}\right) X(mT)\right]$$

$$= R(t - mT) - \sum_{n=-\infty}^{\infty} R(nT - mT) \frac{\sin(\omega_c t - n\pi)}{(\omega_c t - n\pi)} = 0 \qquad (3.59)$$

where it is left to the reader to verify that:

$$R(t - mT) = \sum_{n=-\infty}^{\infty} R(nT - mT) \frac{\sin(\omega_c t - n\pi)}{(\omega_c t - n\pi)}$$

which is shown by substituting $t - mT$ for τ in Eq. (3.57).

Now, utilizing the identity

$$R(t) = \sum_{n=-\infty}^{\infty} R(nT - mT) \frac{\sin[\omega_c(t + mT) - n\pi]}{\omega_c(t + mT) - n\pi}$$

(where this identity is proven by changing $t - mT$ to t in the preceding equation), we now get:

$$E\left[\left(X(t) - \sum_{n=-\infty}^{\infty} X(nT) \frac{\sin(\omega_c t - n\pi)}{(\omega_c t - n\pi)}\right) X(t)\right] = 0 \qquad (3.60)$$

Thus, utilizing Eqs. (3.59) and (3.60), it is easy to show that:

$$E\left[\left(X(t) - \sum_{n=-\infty}^{\infty} X(nT) \frac{\sin(\omega_c t - n\pi)}{(\omega_c t - n\pi)}\right)^2\right] =$$

$$E\left[\left(X(t) - \sum_{n=-\infty}^{\infty} X(nT) \frac{\sin(\omega_c t - n\pi)}{(\omega_c t - n\pi)}\right) X(t)\right] -$$

$$E\left[\left(X(t) - \sum_{n=-\infty}^{\infty} X(nT) \frac{\sin(\omega_c t - n\pi)}{(\omega_c t - n\pi)}\right) X(t)\right] = 0 - 0 = 0$$

3.8 SUMMARY OF SOME USEFUL RESULTS

In what follows, we shall summarize some significant properties concerning complex stationary (wide-sense) processes $X(t)$, $Y(t)$, and $Z(t)$.

(1) $R_X(0) = E[|X(t)|^2]$.

(2) $R_X^*(\tau) = R_X(-\tau)$.

If X is a real process, then

$$R_X(\tau) = R_X(-\tau)$$

(3) If $Z(t) = X(t) + Y(t)$, then

$$R_Z(\tau) = R_X(\tau) + R_Y(\tau) + R_{XY}(\tau) + R_{YX}(\tau)$$

where $R_{XY}^*(\tau) = R_{YX}(-\tau)$

(4) $E[|X(t+\tau) - X(t)|^2] = 2 \operatorname{Re}[R(0) - R(\tau)]$

(5) Assume $R_X(\tau)$ is not periodic, then

$$\lim_{|\tau| \to \infty} R_X(\tau) = |m|^2$$

where $m = E[X]$ and if $X(t)$ and $X(t + \tau)$ are uncorrelated as $|\tau| \to \infty$. Thus, if $E[X(t)] = 0$, then

$$\lim_{|\tau| \to \infty} R_X(\tau) = 0$$

(6) $R_X(0) \geqslant |R_X(\tau)|$, for all τ.

(7) $R(\cdot)$ is an autocorrelation function iff it is nonnegative definite.

(8) $R(\cdot)$ is an autocorrelation function iff its Fourier transform $S(\omega) \geqslant 0$, for all ω.

(9) If $X(t)$ is the input of a time-invariant system with the transfer function $H(\omega)$, then the power spectrum of the output $Y(t)$ is given by:

$$S_Y(\omega) = |H(\omega)|^2 \, S_X(\omega)$$

3.9 IDEAL LOW-PASS SIGNALS

We shall define $X(t)$ to be an ideal low-pass process if $S_X(\omega)$ is given by:

$$S_X(\omega) = \begin{cases} K_0, & \text{for } |\omega| < \omega_c \\ 0, & \text{otherwise} \end{cases}$$

Invoking the inverse Fourier transform, one obtains:

$$R_X(\tau) = \mathscr{F}^{-1} S_X(\omega) = \frac{1}{2\pi} \int_{-\omega_c}^{\omega_c} S_X(\omega) \exp(j\omega\tau) \, d\omega = K_0 \frac{\sin \omega_c \tau}{\pi \tau}$$

Now let us show $R_X(\tau)$ as $\tau \to 0$ (we shall denote $R_X(\tau)$, $\tau \to 0$, as $R_X(0)$). Using L'Hospital's rule on the above equation, we get:

$$\lim_{\tau \to 0} R_X(\tau) = \lim_{\tau \to 0} K_0 \frac{\sin \omega_c \tau}{\pi \tau} = \frac{0}{0} = K_0 \lim_{\tau \to 0} \frac{\frac{d}{d\tau}(\sin \omega_c \tau)}{\frac{d}{d\tau}(\pi \tau)} = K_0 \frac{\omega_c}{\pi}$$

Hence, we can write:

$$R(\tau) = R(0) \frac{\sin \omega_c \tau}{\omega_c \tau} \tag{3.61}$$

From the above equation it is easy to verify that $R(nT) = 0$ for all $n \neq 0$. We can also show that $X(nT)$ processes are mutually orthogonal. This is true since

$$E[X(nT) X(mT)] = R_X[(n-m)T] = 0, \text{ for all } n \neq m$$

Now we shall summarize a significant result via the following theorem:

Theorem 4

A *band-limited* process $X(t)$ is low pass iff $X(nT)$ are mutually orthogonal.

Proof

We have shown that if $X(\cdot)$ is low-pass (characterized by Eq. 3.61), then $X(nT)$ are mutually orthogonal. If the processes $X(nT)$ are mutually orthogonal, all we need to show is Eq. (3.61). Now $R_X(nT)$ by definition is given by:

$$R_X(nT) = E[X(nT) X(0)] = 0, \text{ for all } n \neq 0$$

because of orthogonality. Invoking the sampling theorem (see Eq. 3.57), we get:

$$R(\tau) = \sum_{-\infty}^{\infty} R(nT) \frac{\sin(\omega_c \tau - n\pi)}{(\omega_c \tau - n\pi)}$$

$$= \ldots + 0 + 0 + \ldots R(0) \frac{\sin \omega_c \tau}{\omega_c \tau} + 0 + 0 \ldots = R(0) \frac{\sin \omega_c \tau}{\omega_c \tau}$$

3.10 REPRESENTATION OF BAND-PASS PROCESSES

A signal $X(t)$ whose power spectrum is defined only over a band

$$\omega_0 - \omega_c < |\omega| < \omega_0 + \omega_c$$

and is zero outside the band (see sketch) is called a *band-pass process*. Note that the power spectrum $S_X(\omega)$ is defined only for stationary processes. We observe that the band-pass corresponding to the stationary process $X(t)$ is $2\omega_c$ and is centered at $\omega = \omega_0$.

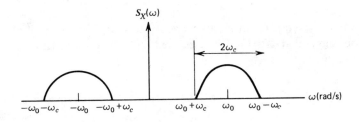

In what follows we shall show that a band-pass process consists of two components, given by:

$$X(t) = X_1(t) \cos \omega t + X_2(t) \sin \omega t \tag{3.62}$$

where $X_1(t)$ and $X_2(t)$ are stationary (wide sense), and $S_{X_1}(\omega) = S_{X_2}(\omega)$. In addition, these power spectrums are shown to be related to $S_X(\omega)$ by the equation:

$$S_{X_1}(\omega) = S_{X_2}(\omega) = \begin{cases} S_X(\omega + \omega_0) + S_X(\omega - \omega_0), & \text{for } |\omega| < 2\omega_c \\ 0, & \text{for } |\omega| > 2\omega_c \end{cases}$$

$$\tag{3.63}$$

We can also show that $S_{X_1 X_2}(\omega)$ and $S_{X_2 X_1}(\omega)$ are related by:

$$S_{X_1 X_2}(\omega) = -S_{X_2 X_1}(\omega) = \begin{cases} j[S_X(\omega - \omega_0) - S_X(\omega + \omega_0)], & \text{for } |\omega| < 2\omega_c \\ 0, & \text{for } |\omega| > 2\omega_c \end{cases}$$

(3.64)

Note that $S_{X_1 X_2}(\omega)$ is not necessarily nonnegative because $R_{X_1 X_2}(\tau)$ is not necessarily nonnegative definite. Furthermore, as a consequence of Eq. (3.51) it can be shown that:

$$E|X(t)|^2 = E|X_1(t)|^2 = E|X_2(t)|^2 \qquad (3.65)$$

Summarizing the above via a theorem is now appropriate; see references [5] and [8].

Theorem 5

$X(t)$ is a band-pass process (implies $X(t)$ is stationary) with the corresponding $S_X(\omega)$ given above (also see accompanying sketch) iff $X(t)$ can be described in Eq. (3.62) and Eqs. (3.63) and (3.64) are satisfied.

Proof

Let $Z(t)$ be a random variable such that $S_Z(\omega) = 4 S_X(\omega)$ and be zero for $\omega < 0$, i.e.,

$$S_Z(\omega) = 4 S_X(\omega) \, 1(\omega) \qquad (3.66)$$

where $1(\cdot)$ denotes the unit step. From Eq. (3.66), we can model $Z(t)$ as the output of a linear system, with the input $X(t)$ and the transfer function $H_1(\omega)$ given by:

$$H_1(\omega) = 2 \, 1(\omega) \qquad (3.67)$$

It is easy to observe that:

$$2 \, 1(\omega) = 1 + \text{sgn } \omega$$

where

$$\text{sgn } \omega = \begin{cases} 1, & \omega > 0 \\ -1, & \omega < 0 \end{cases}$$

Thus, $Z(t)$ can be modeled by an alternate approach, i.e.,

$$Z(t) = X(t) + j\check{X}(t) \tag{3.68}$$

where $\check{X}(t)$ is defined by using $X(t)$ as the input of a linear system with a corresponding transfer function $H(\omega)$ given by:

$$H(\omega) = -j \text{ sgn } \omega, \text{ i.e., } h(t) = \frac{1}{\pi t} \tag{3.69}$$

Hence,

$$\check{X}(t) = \int_{-\infty}^{\infty} h(t-\tau) X(\tau) \, d\tau = \frac{1}{\pi} \int_{-\infty}^{\infty} \frac{X(\tau)}{t-\tau} \, d\tau = \frac{1}{\pi t} * X(t)$$

$$\tag{3.70}$$

We define $\check{X}(t)$ given by Eq. (3.70) as the Hilbert transform of $X(t)$. The process $Z(t)$ is called the analytic signal associated with $X(t)$. It is useful to observe that if $X(t)$ is the input with the transfer function $H(\omega) = -j \text{ sgn } \omega$, then the output is $\check{X}(t)$ because:

$$(H(\omega))^2 = (-j \text{ sgn } \omega)^2 = -1 \tag{3.71}$$

From Eq. (3.71), we can verify $|H(\omega)|^2 = 1$ and

$$S_{\hat{X}}(\omega) = S_X(\omega) \text{ and } R_{\hat{X}}(\tau) = R_X(\tau) \tag{3.72}$$

Let $\widetilde{\widetilde{X}}$ denote the output of a system with the input $\widetilde{X}(t)$ and the transfer function $H(\omega) = -j\ \text{sgn}\ \omega$; then it is easy to verify that:

$$\widetilde{\widetilde{X}}(t) = -X(t) \tag{3.73}$$

Hence, for the processes $X(t)$, $\widetilde{X}(t)$, and $\widetilde{\widetilde{X}}(t)$, their behavior can be summarized as:

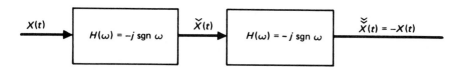

Now, utilizing the facts that:

$$S_{X\widetilde{X}}(\omega) = S_{XX}(\omega)\, H^*(\omega) = j\ \text{sgn}\ \omega\, S_X(\omega) \tag{3.74}$$

and

$$S_{\widetilde{X}X}(\omega) = S_{XX}\, H(\omega) = -j\ \text{sgn}\ \omega\, S_X(\omega) \tag{3.75}$$

then

$$S_{X\widetilde{X}}(\omega) = -S_{\widetilde{X}X}(\omega) \tag{3.76}$$

and

$$R_{X\widetilde{X}}(\tau) = -R_{\widetilde{X}X}(\tau) \tag{3.77}$$

Now we shall consider the process $Z(t)\exp(-j\omega_0 t)$, and let

$$Z(t)\exp(-j\omega_0 t) = X_1(t) - j X_2(t) \tag{3.78}$$

That is,

$$X_1(t) = \text{Re}\,[Z(t)\exp(-j\omega_0 t)] = X(t)\cos\omega_0 t + \widetilde{X}(t)\sin\omega_0 t \tag{3.79}$$

$$X_2(t) = \text{Im}\,[Z(t)\exp(-j\omega_0 t)] = X(t)\sin\omega_0 t - \check{X}(t)\cos\omega_0 t$$

(3.80)

from which we can obtain:

$$X(t) = X_1(t)\cos\omega_0 t + X_2(t)\sin\omega_0 t \qquad (3.81)$$

$$\check{X}(t) = X_1(t)\sin\omega_0 t - X_2\cos\omega_0 t \qquad (3.82)$$

From Eqs. (3.79) and (3.80), we obtain:

$$E[X_1(t+\tau)X_1(t)] = \frac{1}{2}\{[R_X(\tau) + R_{\check{X}}(\tau)]\cos\omega_0\tau$$

$$+ [R_{X\check{X}}(\tau) - R_{\check{X}X}(\tau)]\sin\omega_0\tau$$

$$+ [R_X(\tau) - R_{\check{X}}(\tau)]\cos\omega_0(2t+\tau)$$

$$+ [R_{X\check{X}}(\tau) + R_{\check{X}X}(\tau)]\sin\omega_0(2t+\tau)\}$$

Now if we use Eqs. (3.72) and (3.77) in the above, we obtain:

$$R_{X_1}(\tau) = R_X\cos\omega_0\tau + R_{X\check{X}}(\tau) \qquad (3.83)$$

which is stationary, and, similarly,

$$R_{X_2}(\tau) = R_X\cos\omega_0\tau + R_{X\check{X}}(\tau)\sin\omega_0\tau \qquad (3.84)$$

which implies:

$$R_{X_1}(\tau) = R_{X_2}(\tau) \qquad (3.85)$$

Now, $S_{X_1}(\omega)$ can be obtained from (3.83) by:

$$S_{X_1}(\omega) = \frac{1}{2}[S_X(\omega + \omega_0) + S_X(\omega - \omega_0)]$$

$$+ \frac{1}{2}[S_X(\omega + \omega_0) \operatorname{sgn}(\omega + \omega_0) - S_X(\omega - \omega_0) \operatorname{sgn}(\omega - \omega_0)]$$

(3.86)

Let $S_q(\omega)$ denote $S_X(\omega)$ where we translate $S_X(\omega)$ from its center at ω_0 to the zero frequency. It can be verified that:

$$S_X(\omega + \omega_0) = S_q(\omega) + S_q(-\omega - 2\omega_0) \tag{3.87}$$

$$S_X(\omega - \omega_0) = S_q(-\omega_0) + S_q(\omega - 2\omega_0) \tag{3.88}$$

substituting (3.87) and (3.88) into (3.86) yields:

$$S_{X_1}(\omega) = S_q(\omega) + S_q(-\omega) \tag{3.89}$$

and, further, it can be shown that:

$$S_{X_1}(\omega) = S_q(\omega) + S_q(-\omega) = \begin{cases} S_X(\omega + \omega_0) + S_X(\omega - \omega_0), & |\omega| < 2\omega_c \\ \\ 0, & |\omega| > 2\omega_c \end{cases}$$

(3.90)

From Eq. (3.85), we also have:

$$S_{X_1}(\omega) = S_{X_2}(\omega)$$

Hence, we have shown (see Eq. 3.90) that $X_1(t)$ and $X_2(t)$ are low-pass processes.

To find $R_{X_1 X_2}(\tau)$, which is equal to $-R_{X_2 X_1}(\tau)$, we use Eqs. (3.79) and (3.80) to obtain:

$$R_{X_1 X_2}(\tau) = -R_{X_2 X_1}(\tau) = R_X(\tau) \sin \omega_0 \tau - R_{X\check{X}}(\tau) \cos \omega_0 \tau$$

and

$$S_{X_1 X_2}(\omega) = -S_{X_2 X_1}(\omega) = j[S_q(-\omega) - S_q(\omega)]$$

$$= \begin{cases} j[S_X(\omega - \omega_0) - S_X(\omega + \omega_0)], & |\omega| < 2\omega_c \\ 0, & |\omega| > 2\omega_c \end{cases}$$

(3.91)

It is easy to verify that:

$$E(|X_1(t)|^2) = E(|X_2(t)|^2)$$

because $S_{X_1}(\omega) = S_{X_2}(\omega)$. Similarly, it is easy to verify that:

$$E(|X(t)|^2) = E(|X_1(t)|^2) = E(|X_2(t)|^2)$$

The representation given by (3.81) and (3.82) of $X(t)$ and $\check{X}(t)$ is known as the quadrature component representations.

We should point out that the band-limited processes have wide ranges of applications in many engineering disciplines, especially in communication systems. The object of many communication systems is to accurately transmit a message from one point to another. Very often, a message has a low-pass power spectrum. A typical example of such a message is a video signal. A very large antenna would be required to transmit such a message because of its low-frequency characteristics. To obviate the need for such an antenna, and thus to utilize the transmission media more effectively, a band-pass transmission at a much higher frequency is required. This procedure is called modulation. At the receiving terminal we must undo the modulation process (demodulation) to recover the original low-pass signal from band-pass.

EXERCISES

3.1 In an RC circuit, where $R = 1\,\Omega$ and $C = 1\,\text{F}$, let the input voltage source be a random process $X(t)$ such that $S_X(\omega) = K_0$ and the output be the voltage across the capacitor denoted by $Y(t)$. Assume that $X(t)$ and $X(t+\tau)$ are uncorrelated as $|\tau| \to \infty$, then:

(a) Show the transfer function $H(j\omega)$ is given by:

$$H(j\omega) = \frac{1}{1+j\omega}$$

(b) Obtain $S_Y(\omega)$

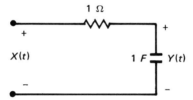

(c) If $R_X(\tau) = m_X^2$, find the mean and the variance of $Y(t)$.

(d) Obtain the variance of $Y(t)$ and comment on your result as $|\tau| \to \infty$.

3.2 Let $Y(t)$ be a process given by:

$$Y(t) = X(t+1) - X(t-1)$$

where $X(t)$ is a zero mean stationary random variable. Show that:

$$S_Y(\omega) = 4\,S_X \sin^2 \omega$$

3.3 Determine the correlation function of the white noise $S(\omega)$ given by:

$$S(\omega) = \begin{cases} \mathcal{N}, & \omega_1 < |\omega| < \omega_2 \\ 0, & \text{otherwise} \end{cases}$$

3.4 Repeat the previous problem for:

$$S(\omega) = \begin{cases} \mathcal{N}, & |\omega| < \omega_c \\ 0, & \text{otherwise} \end{cases}$$

3.5 Determine the correlation function of the process $X(t)$ with its power spectrum given by:

$$S(\omega) = \frac{1}{(4 + \omega^2)^2}$$

3.6 In Problem 2.9, obtain $S_X(\omega)$ and $S_Y(\omega)$.

3.7 The input $X(t)$ to a linear time-invariant system has the correlation function $R_X(\tau) = \delta(\tau)$. Assume the output is $Y(t)$. Then find $R_Y(\tau)$ and $R_{XY}(\tau)$ as well as their corresponding power spectrums, given:

(a) $h(t) = 1$, given $0 < t < T$ and zero otherwise.

(b) $h(t) = t \exp(-2t)$, $t > 0$.

3.8 Let a white noise process have power spectrum of $\mathcal{N}$. Assume this process is transmitted through a band-pass filter, shown in the figure below. Then perform the following:

(a) Justify whether or not the output of the filter is band-pass.

(b) Determine the power spectrum of the output as well as its mean-square value.

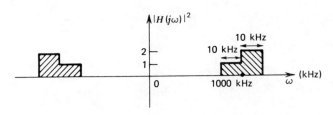

3.9 Let $n(t)$ be a white noise band-pass process given by the figure below.

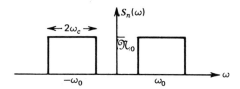

We know the band-pass $n(t)$ will consist of two components such that:

$$n(t) = n_1(t) \cos \omega t + n_2(t) \sin \omega t$$

where $n_1(t)$ and $n_2(t)$ are wide-sense stationary, and $S_{n_1}(\omega) = S_{n_2}(\omega)$.

(a) Obtain $S_{n_1}(\omega)$ and $S_{n_1 n_2}(\omega)$.

(b) Determine $E[n^2(t)]$.

CHAPTER 4
ESTIMATION THEORY

4.1 INTRODUCTION

Heuristically speaking, stochastic estimation is the operation of assigning a value to an unknown parameter based on contaminated (noisy) observations or measurements involving some function of the parameter. The noise contaminating the uncontaminated signal is assumed to have known statistical properties. The assigned value is called an estimate and the system or functions yielding the estimate is called the estimator. In many applications it is meaningful to assign a cost to an estimate representing a quantitative measure of how "good" an estimate is. This cost function should be a function of estimation errors, i.e., the difference between the true value and the estimated value. An optimal estimate is a function of received observations (measurements) which is chosen to minimize the expected value of the cost function. An estimator yielding such an optimal estimate is called a Bayes estimator. A basic feature of the Bayes estimator is that it requires a knowledge of an *a priori* probability density function.

The present-day theories of estimation in the time domain, with few exceptions, owe their creation to Wiener and Kolmogrov. They basically considered the problem of "optimal" separation of a signal $s(t)$ which was contaminated by additive noise $n(t)$. We denote the contaminated signal $Y(t)$ and call it observation, i.e.,

$$Y(t) = s(t) + n(t)$$

We shall use the same notation for the signal whether it is a process or ensemble throughout this chapter.

Wiener studied the continuous-time problems and assumed that $s(t)$ and $n(t)$ were typical numbers drawn from ensembles of those functions which were wide-sense stationary with known first two moments. In addition, he assumed the availability of a semi-infinite observation and solved the problem of linear least square estimation, reducing it to the problem of solving a very difficult integral equation, the so-called "Wiener Hopf equation." That is, the optimal solution by Wiener's method would terminate with an integral equation whose solution would be needed to separate $s(t)$ optimally from the noise.

Even if one is willing to accept physically that the signal and noise be stationary and the observation be given over a semi-infinite interval, there remains a major problem: computation of optimal solutions which utilizes the "Wiener-Hopf integral equation," where its solution with the exception of some academic problems is extremely complicated and computationally infeasible. The statistical assumptions are also very stringent, which further limits the applicability to many practical problems such as those in orbital mechanics, space tracking, and countless others.

Kalman and Bucy revived estimation theory. They provided an alternative method to that of Wiener by assuming the availability of the observation over a finite interval and not limiting themselves to stationary processes. Kalman and Bucy considered the special class of processes which could be generated by a white noise forcing function serving as the input to a finite dimensional dynamic system (explained in the following sections). They assumed complete knowledge about the model in order to avoid certain very difficult problems.

The primary interest in Kalman's estimation technique is in practical applications. We shall first discuss some basic results of mean-square estimation (quadratic mean) via the classical approach as well as some basic results of mean square estimation via Kalman-Bucy filtering. The latter involves the solution of the so-called "state estimation problems" associated with finite-dimensional linear dynamic systems operating in a stochastic environment. A discussion of characterization of linear systems via the state variable approach will be carried out later in the chapter.

4.2 SYSTEMS AND MODELING

Physical systems are normally characterized by models consisting of idealized elements which can be defined mathematically. Choosing an appropriate model which characterizes all the features of the physical system is very important and generally very difficult. For example, if an unnecessarily complicated model is used, it may be impossible to analyze the model. On the other hand, if an extremely simple model is utilized, the results obtained by it may not be a realistic approximation to the physical phenomenon. Generally speaking, a model of the physical system may be mathematically expressed via integro-differential equations. Although in real life very few systems are linear, they can often be adequately approximated by linear models over an operating range of interest. The treatment of a nonlinear system is extremely difficult; therefore, it is often necessary to assume that the system under study is a linear system. The general steps involved in the study of a physical system may be described by Figure 4-1.

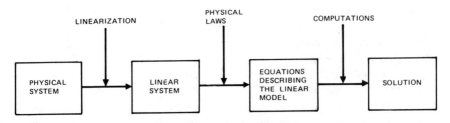

Fig. 4-1. Characterization of a Physical System.

A convenient method of characterizing a linear system is by its input-output relationship. In general, a system may have many inputs and many outputs.

The electric circuit given by Figure 4-2 can be considered as a system with a single input and a single output, where $e(t)$ is the input and $e_0(t)$ is the output.

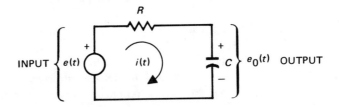

Fig. 4-2. RC Electric Circuit

In a linear system, the variables $u(t)$ and $y(t)$ can be related by

$$y(t) = \int_{t_0}^{t} h(t, \lambda) \, u(s) \, d\lambda, \quad u(t_0) = 0$$

if the system is causal and is at rest at t_0, where $h(t, \lambda)$ is called the system's impulse response. If the system is characterized by a constant coefficient differential equation, then it can be shown that $h(t, \lambda) = h(t - \lambda)$.

4.2.1 State Variable Characterization of a Linear System

The classical method of describing a linear system is by its impulse response and, if the system is also time-invariant, by its frequency domain transfer function. It should be emphasized that frequency domain analysis, although the most attractive, can only be utilized for time-invariant linear systems. In nonlinear and time-varying linear systems, the frequency domain analysis cannot be utilized to advantage. Even in the time-invariant case the frequency domain transfer function suffers from the major disadvantage that all the initial conditions of the system are ignored. The analysis and the synthesis of linear systems, time-varying or not, is a formidable task for multivariable systems (vector input-output), and determining the interrelated effects in a multivariable system is a complicated and exhausting process.

The modern alternative to classical methods of describing a system is by the "state variable" technique, which is a matrix method for handling multivariable systems. The technique aids conceptual thinking and provides a unifying basis for quantitative information about the system. The state of the system is defined in terms of a minimal set of variables $X_1(t), \ldots, X_2(t), \ldots, X_n(t)$, such that information about these variables at time $t = t_0$ along with the input $u(t)$ for all $t \geqslant t_0$ uniquely determines the output $Y(t)$ for $t \geqslant t_0$.

The state is the answer to the following question: "Suppose $u(t)$ for $t \geqslant t_0$ is known. What additional information is needed to completely obtain $Y(t)$ for $t \geqslant t_0$?" We shall discuss the concept of state later in the chapter and give examples of its use.

4.3 MEAN-SQUARE ESTIMATION

In this section we shall construct a mean-square performance index in order to carry out the estimation process. Throughout this section, unless specified otherwise, the norm of a random vector X is defined as

$$\|X\|^2 = X'X$$

where X is a column vector, and prime denotes the transpose.

Now let us specify the estimation problem. Let two random vectors X and Y of dimensions n and m, respectively, be jointly distributed. Suppose Y is a measurement which in general has been contaminated by noise. It is intuitively obvious that the received measurement, Y, should improve the information about X. That is, if we had an *a priori* guess about X, knowledge of Y should improve the information about X. To be more specific, let us ask ourselves the question, "Given the measurement $Y = y$, what is the best estimate of X, denoted as $\hat{X}(Y)$, corresponding to the random vector X?" The concept of "best" has not been defined, but the most popular criterion is the mean-square estimate. Thus we are seeking to obtain the estimate, $\hat{X}(Y)$, which is the function of measurement $Y = y$ such that:

$$E[\|X - \hat{X}(Y)\|^2 |_{Y=y}] = \min E[\| X - l \|^2 |_{Y=y}] \qquad (4.1)$$

over all random vectors l.

The criterion given by (4.1) is referred to in the literature by the following names:

(1) Minimum mean square estimate

(2) Least square estimate

The solution of (4.1) is relatively simple and is given by:

$$\hat{X}(y) = E[X|Y] \qquad (4.2)$$

Hence, we are assuming a cost function associated with the uncertainty of X. We choose $\hat{X}(y)$ as the best estimate that Y has the value y under condition (4.1) and claim it is given by condition (4.2).

Let us verify (4.2).

$$E[\|X - l\|^2|_Y] = E[(X - l)'(X - l)|_Y]$$

$$= E[(X'X - l'X - X'l + l'l)|_Y]$$

$$= E[\|l - E[X|Y]\|^2] + E[\|X\|^2|Y] - \|E[X|Y]\|^2$$

From the above equation, the only term that has l involved in it is the first term, and the right-hand side of the above equation is minimum if and only if $E[\| l - E[X|Y]\|^2] = 0$, which implies that the best solution of l, is:

$$\hat{l} = E[X|Y] = \hat{X}(Y) \qquad (4.3)$$

It is very important to mention that, in general, $\hat{X}(Y)$ is a random vector, since $\hat{X}(\cdot)$ is a function of the random vector Y. However, for each measurement $Y = y$, the corresponding $\hat{X}(y)$ is a deterministic outcome of that random vector.

Let g be a function of Y from $R^m \to R^n$ and assume $f_Y(y) \neq 0$. From condition (4.1) it is obvious that:

$$E[\| X - \hat{X}(Y) \|^2 |_Y] \leq E[\| X - g(Y) \|^2 |_Y] \tag{4.4}$$

because we substitute $g(Y)$ for l in that equation.

Now, let us take the expected value of both sides of Eq. (4.4):

$$E(E[\| X - \hat{X}(Y) \|^2 |_Y]) \leq E(E[\| X - g(Y) \|^2 |_Y]) \tag{4.5}$$

Utilizing the identities:*

$$E(E[\| X - \hat{X}(Y) \|^2 |_Y]) = E[\| X - \hat{X}(Y) \|^2]$$

and

$$E(E[\| X - g(Y) \|^2 |_Y]) = E[\| X - g(Y) \|^2]$$

we obtain:

$$E[\| X - \hat{X}(Y) \|^2] \leq E[\| X - g(Y) \|^2] \tag{4.6}$$

Equation (4.6) states a very significant result: the estimate $\hat{X} = E[X|Y]$ is the best solution for the unconstrained case. Thus, the result can be appropriately summarized via a theorem.

Theorem 1

For two jointly distributed random vectors X and Y with joint probability density functions $f_{XY}(x,y)$ and $f_Y(y) \neq 0$, the best estimate of $E[\| X - g(Y) \|^2]$ is given by:

$$\hat{X}(Y) = E[X|Y] \tag{4.7}$$

---

*We are using the general result $E(E[h(X,Y)|_Y]) = E[h(X,Y)]$.

Remark 1. If $e = X - \hat{X}$, then $\hat{X} = E(X|Y)$ is uncorrelated with any mapping of the random vector Y. Mathematically we can write:

$$E[g(Y)e'] = 0$$

where the prime denotes the transpose. The reader is advised to verify this equation.

4.4 LINEAR ESTIMATE

The estimate just obtained is indeed the best with respect to the mean-square cost function. However, $\hat{X}(Y)$ is a nonlinear function of Y (for the general case), and it is extremely difficult to obtain the exact relationship. Since very often $f_{XY}(x, y)$ is not available, then $E(X|Y)$ may not be achievable either.

Now we shall do the next best thing and introduce a constraint that $\hat{X}(Y)$ has a linear form of Y. That is,

$$\hat{X} = AY + \mathbf{b} \tag{4.8}$$

where A is an $n \times m$ matrix and $\mathbf{b}$ is an n-vector. With the constraint (4.8) on Eq. (4.7), we get:

$$E[\| X - AY - \mathbf{b}\|^2] = E[(X - AY - \mathbf{b})'(X - AY - \mathbf{b})] \tag{4.9}$$

Now we can choose A and $\mathbf{b}$ (parameters) such that Eq. (4.9) is minimized. Let us denote the optimal values of A and $\mathbf{b}$ as A_0 and $\mathbf{b}_0$, thus $\hat{X}(y)$ shall be given by:

$$\hat{X}(Y) = A_0 Y + \mathbf{b}_0 \tag{4.10}$$

Without any loss of generality, assume that X and Y have zero mean. To minimize the cost function given by Eq. (4.9), we shall calculate A_0 and $\mathbf{b}_0$ in the usual manner by setting:

$$\frac{\partial}{\partial \mathbf{b}} E[\| X - AY - \mathbf{b}\|^2] = \frac{\partial}{\partial \mathbf{b}} E[(X - AY - \mathbf{b})'(X - AY - \mathbf{b})] = 0$$

and

$$\frac{\partial}{\partial A} E[\|X - AY - \mathbf{b}\|^2] = \frac{\partial}{\partial A} E[(X - AY - \mathbf{b})'(X - AY - \mathbf{b})] = 0$$

From the first equation, we find:

$$\mathbf{b}_0 = 0$$

and, from the second,

$$A_0 = E(XY')[E(YY')]^{-1} = C_{XY} C_Y^{-1} \qquad (4.11)$$

since X and Y are zero mean. Hence,

$$\widehat{X}(Y) = C_{XY} C_Y^{-1} Y \qquad (4.12)$$

Now, if X and Y do not have zero mean, the random variables $X - m_X$ and $Y - m_Y$ have zero means. Applying (4.12) yields:

$$\widehat{X - m_X} = C_{XY} C_Y^{-1} (Y - m_Y)$$

or, equivalently,

$$\widehat{X}(Y) = m_X + C_{XY} C_Y^{-1} (Y - m_Y) \qquad (4.13)$$

In the next section, we shall show that the best estimate can be derived by a different approach, the so-called "orthogonality principle." The orthogonality principle is one of the most important ideas in linear estimation theory. Let us define an important concept.

Definition 1

An estimate $\widehat{X}(Y)$ is defined to be a conditional unbiased estimate if:

$$E\widehat{X}(Y) = X \qquad (4.14)$$

That is, the average (with respect to $f_Y(y)$) of the estimate is equal to the true value. This definition is motivated by the fact that if we are receiving a perfect measurement Y (i.e., Y is not random), then $\hat{X}(Y)$ is not a random variable, and

$$E_Y \hat{X}(Y) = \hat{X}(Y) = X$$

That is, if there were no measurement errors, and thus no uncertainty, then the estimate is identical to the true value. Also for the unbiased estimate, we can write:

$$E[(X - \hat{X})(X - \hat{X})'] = E[(\hat{X} - E\hat{X})(\hat{X} - E\hat{X})'] = C_{\hat{X}} = E(ee') \tag{4.15}$$

where

$$e = X - \hat{X}$$

4.5 ORTHOGONALITY PRINCIPLE

In this section we shall assume, without loss of generality, that all parameters are of zero mean, unless specified otherwise. For example, if the mean of X is non-zero, then we shall introduce a new random variable $\overline{X} = X - m_X$ which will have zero mean (as in the previous section).

The concept of orthogonality is extremely important in the theory of linear mean-square estimation. We shall show that the orthogonality principle will serve as a necessary and sufficient condition that the linear estimate $\hat{X}$ be the best. The orthogonality principle states that if the measurement Y is orthogonal to the error $e = X - \hat{X}$, i.e.,

$$E[(X - \hat{X}) Y'] = E[eY'] = 0 \tag{4.16}$$

then the estimate $\hat{X}$ is the best linear m.s.e.

Definition 2

An estimate $\hat{X}$ is optimal if it is the best linear mean-square estimate.

Before proving the result given by Eq. (4.16), it is recommended that the reader thoroughly review Appendix B, especially the discussion of the particular Hilbert space denoted by H.

4.5.1 Discussion of Vector Spaces

In Appendix B an infinite-dimensional vector space, which is a particular Hilbert space, is defined. Basically, H is generated by the set of all random variables X such that

$$EX = 0 \quad \text{and} \quad E|X|^2 < \infty$$

and the random variables X_1 and X_2 are equivalent if $E|X_1 - X_2|^2 = 0$.

We can assume without any loss of generality that the random variables are real. Thus, using Eqs. (B.18) and (B.19) of Appendix B, we get:

$$(X, Y) = E(XY) \qquad (4.17)$$

and

$$\|X\|_{q.m.} = (X, X)^{1/2} = (E|X|^2)^{1/2} \qquad (4.18)$$

As in Appendix B, let M denote the subspace of H generated by $X_1, X_2, \ldots, X_n$ assumed to be linearly independent. We know H can be decomposed into the direct sum of M and $M^\perp$:

$$H = M \oplus M^\perp$$

That is, every vector $X \in H$ is given by:

$$X = \eta_1 + \eta_2$$

where

$$\eta_1 \in M \text{ and } \eta_2 \in M^\perp.$$

Recall the projection of X denoted as P on M is given by:

$$PX = \eta_1$$

and the projection of X denoted by Q on $M^\perp$ is given by:

$$QX = \eta_2$$

where

$$P + Q = I$$

and I is the identity operator. Hence,

$$Q = I - P \tag{4.19}$$

We can now use the concept of a vector space to obtain a significant result.

Theorem 2

Let X be a variable $\in H$, and let Z be a vector $\in M$. Then

$$\|X - Z\|_{q.m.}^2 = E[(X - Z)'(X - Z)]$$

reaches its minimum if and only if

$$Z = PX$$

Proof

For any $X \in H$, we have

$$X = \eta_1 + \eta_2$$

where $\eta_1 \in M$, $\eta_2 \in M^\perp$, and $\eta_1 = PX$, $\eta_2 = (I - P)X$. We also have:

$$\|X - Z\|_{q.m.}^2 = E[(X - Z)'(X - Z)]$$

$$= E\{[(X - \eta_1) + (\eta_1 - Z)]'$$
$$[(x - \eta_1) + (\eta_1 - Z)]\} \tag{4.20}$$

In the above equation $X - \eta_1$ is orthogonal to M, i.e., $X - \eta_1 \in M^\perp$, while η_1, Z, $\eta_1 - Z$ are all members of M. Utilizing these facts in Eq. (4.20) yields:

$$\|X - Z\|^2_{q.m.} = \|X - \eta_1\|^2_{q.m.} + \|\eta_1 - Z\|^2_{q.m.} \quad (4.21)$$

From the above equation, it is obvious that:

$$\|X - Z\|^2_{q.m.} \geq \|X - \eta_1\|^2_{q.m.} \quad (4.22)$$

since $\|\eta_1 - Z\|^2_{q.m.} \geq 0$. Thus, the inequality in (4.22) becomes an equality if and only if

$$Z = \eta_1 = PX$$

4.5.2 Application of Theorem 2

Assume that we have received m measurements that are linearly independent, say, the random vectors $Y_1, Y_2, \ldots, Y_m$. Let M be the vector space spanned (generated) by the set of all linear combinations of $Y_1, \ldots, Y_m$. According to the theorem, $\|X - Z\|^2_{q.m.}$ is minimized if and only if

$$Z = PX \in M$$

If $Z \in M$, then Z can be written as the linear combination of $Y_1, Y_2, \ldots, Y_m$.

Claim 1. Let $Y_1, Y_2, \ldots, Y_m$ be the measurement vectors (observations), and let M denote the vector space generated by these measurement vectors. Then vector $\hat{X}$ is an optimal estimate of X if and only if $\hat{X}$ is the projection of X onto M.

Claim 2. The vector $\hat{X}$ is an optimal estimate of X if and only if the error $e = X - \hat{X}$ is orthogonal to the observation vectors $Y_1, Y_2, \ldots, Y_m$, i.e.,

$$E[(X - \hat{X}) Y'_i] = E[e\, Y'_i] = 0, \text{ for } i = 1, \ldots, m$$

Claim 2 follows from Claim 1, because if $\hat{X}$ is the projection of X onto M, then $X - \hat{X} \in M^\perp$.

Example 1

In Section 4.3, we derived the optimal estimate $\hat{X}$ as:

$$\hat{X} = C_{XY} C_Y^{-1} Y$$

when we had one observation vector only (see Eq. 4.12). Use the orthogonality principle to derive the same result.

Solution

$$E(eY') = E[(X - \hat{X}) Y'] = 0$$

Since we know $\hat{X} = AY$, where A is to be determined, then

$$E[(X - AY) Y'] = E[XY' - AYY'] = 0$$

This is true if and only if

$$E(XY') = C_{XY} = AE(YY') = AC_Y$$

Assuming that the inverse C_Y^{-1} exists, then it is trivial to see

$$A = C_{XY} C_Y^{-1}$$

as asserted.

Example 2

Let both X and Y be random variables such that:

$$m_n = E(Y^n) \text{ and } E(Y) = 0$$

Show the best linear m.s.e. of $X = Y^2$ is given by:

$$\hat{X} = \frac{m_3 Y}{m_2} + m_2$$

Solution

We know the mean of the value X is:

$$m_X = E(X) = E(Y^2) \neq 0$$

Thus, our estimate $\hat{X}$ shall have the form:

$$\hat{X} = aY + b$$

where we should minimize

$$E[X - (aY + b)]^2$$

with respect to a and b as in Section 4.3. However, this approach is relatively lengthy.

Using the orthogonality principle, the solution is much more direct. Let Z be defined such that:

$$Z = X - E(X) = Y^2 - m_2 \qquad (4.23)$$

Z has zero mean, since $E(Z) = EX - EX = 0$. Now we can use the orthogonality principle:

$$E[(Z - \hat{Z})Y] = 0 \qquad (4.24)$$

where

$$\hat{Z} = AY \qquad (4.25)$$

Hence, from (4.24) and (4.25),

$$E[(Z - AY)Y] = 0$$

which implies:

$$A = \frac{E(ZY)}{E(Y^2)} \qquad (4.26)$$

Substituting Z from Eq. (4.23), we get:

$$A = \frac{E[(Y^2 - m_2)Y]}{m_2} = \frac{m_3 - m_2 E(Y)}{m_2} = \frac{m_3}{m_2}$$

Hence,

$$\hat{Z} = \hat{X} - m_2 = AY = \frac{m_3}{m_2} Y$$

From the above, it is obvious that:

$$\hat{X} = \frac{m_3}{m_2} Y + m_2$$

4.6 LINEAR MEAN-SQUARE ESTIMATE OF CONTINUOUS STOCHASTIC SIGNALS

As discussed in the introduction, Wiener and Kolmogorov formulated the problem of optimal separation of signal $s(t)$ from noise $n(t)$, where the continuous measurement $Y(t)$ is given by:

$$Y(t) = s(t) + n(t)$$

where both $s(t)$ and $n(t)$ are assumed to be wide-sense stationary processes.

We shall use the same notation for the ensemble and the process. The purpose of the Wiener-Kolmogorov (W-K) theory is to extract the signal from the noise, that is, to derive an optimal estimate of $s(t)$ denoted by $\hat{s}(t)$, where the performance index is as before the mean square.

Let us consider a more general case that $s(t)$, namely, $s(t + \alpha)$. Let $\hat{s}(t + \alpha)$ be the corresponding optimal estimate and let the error $e(t + \alpha)$ be defined as:

$$e(t + \alpha) = s(t + \alpha) - \hat{s}(t + \alpha)$$

There are three important cases:

(a) If $\alpha > 0$, then $\hat{s}(t + \alpha)$ is called the (optimal) prediction of $s(t + \alpha)$.

(b) If $\alpha = 0$, then $\hat{s}(t)$ is called the (optimal) filter for $s(t)$.

(c) If $\alpha < 0$, then $\hat{s}(t + \alpha)$ is called (optimal) smoothing of $s(t + \alpha)$.

4.7 THE WIENER-KOLMOGOROV THEORY

The Wiener-Kolmogorov (W-K) theory utilizes the best linear mean-square-estimate criteria applied to stochastic signals in a manner to be specified. The W-K theory emphasizes the time-domain analysis. The smoothing-and-prediction problem was first treated by Wiener and almost simultaneously by Kolmogorov. To make Wiener filtering feasible, some assumptions concerning the signal $s(t)$, the noise $n(t)$, and the measurement

$$Y(t) = s(t) + n(t) \tag{4.27}$$

are made. We shall confine ourselves to one-dimensional signals throughout this section for the sake of simplicity.

Assume that $s(t)$ and $n(t)$ are wide-sense stationary processes of zero mean, such that $s(t)$ and $n(t)$ are uncorrelated, i.e.,

$$E[s(t)\, n(t)] = 0$$

Now let us assume the measurement $Y(t)$ is the input of a linear time-invariant system, characterized via the impulse function $h(t)$ (see sketch).

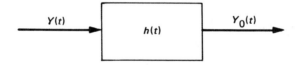

The output signal $Y_0(t)$ can be written as:

$$Y_0(t) = \int_{-\infty}^{\infty} h(\tau)\, Y(t - \tau)\, d\tau \tag{4.28}$$

Note that $Y_0(t)$ is a linear function of $Y(\cdot)$.

Now the objective is to select the appropriate impulse function denoted by $\hat{h}(t)$ such that we minimize the mean square of:

$$E[e_\alpha^2(t)] = E\{[Y_0(t) - s(t + \alpha)]^2\} \tag{4.29}$$

where

$$e_\alpha(t) = Y_0(t) - s(t + \alpha) \tag{4.30}$$

and α is a fixed constant.

The impulse response $\hat{h}(t)$ that minimizes the performance index given by (4.29) gives rise to an optimal solution. The filter with impulse response $\hat{h}(\cdot)$ is called the optimal filter.

$E[e_\alpha^2(t)]$ can be obtained in terms of covariances $s(t)$ and $n(t)$, since

$$E[e_\alpha^2(t)] = E\{[Y_0(t) - s(t + \alpha)]^2\}$$

$$= E[Y_0^2(t)] + E[s^2(t + \alpha)] - 2E[Y_0(t)\, s(t + \alpha)] \tag{4.31}$$

The first term of the above can be written as:

$$E[Y_0^2(t)] = E\left\{\int_{-\infty}^{\infty}\int_{-\infty}^{\infty} h(\tau)\, h(\sigma)\, Y(t - \tau)\, Y(t - \sigma)\, d\tau\, d\sigma\right\} \tag{4.32}$$

Assume that the expected value can operate inside the integrand. Then, utilizing the property of the stationarity, we can verify:

$$E[Y(t - \tau)\, Y(t - \sigma)] = E[Y(\tau)\, Y(\sigma)] = R_Y(\tau - \sigma)$$

$$= E\{[s(\tau) + n(\tau)]\, [s(\sigma) + n(\sigma)]\}$$

$$= R_s(\tau - \sigma) + R_n(\tau - \sigma) \tag{4.33}$$

Hence,

$$E[Y_0^2(t)] = \int_{-\infty}^{\infty}\int_{-\infty}^{\infty} h(\tau)\, h(\sigma)\, [R_s(\tau - \sigma) + R_n(\tau - \sigma)]\, d\tau\, d\sigma$$

$$\tag{4.34}$$

Similarly, we can verify:

$$E[Y_0(t)\, s(t + \alpha)] = \int_{-\infty}^{\infty} h(\tau)\, R_s(\tau + \alpha)\, d\tau \qquad (4.35)$$

and remembering that $R_s(0) = E[s^2(t + \alpha)]$ and substituting this and (4.34) and (4.35) into (4.31) yields:

$$E[e_\alpha^2(t)] = R_s(0) - 2 \int_{-\infty}^{\infty} h(\tau)\, R_s(\tau + \alpha)\, d\tau$$

$$+ \int_{-\infty}^{\infty} \int_{-\infty}^{\infty} h(\tau)\, h(\sigma)\, [R_s(\tau - \sigma) + R_n(\tau - \sigma)]\, d\tau\, d\sigma$$

$$(4.36)$$

The above equation demonstrates that the optimal solution depends on the autocorrelation (covariance) functions only. It should be emphasized that this is an extremely important result, because the optimal filter $h(t)$ is obtained from the knowledge of $R_s(\cdot)$ and $R_n(\cdot)$ and not directly from $s(t)$ and $n(t)$. Hence, there are infinitely many signals that give rise to the optimum solution, all having the same autocorrelation (covariance) function. Wiener minimized $E[e_\alpha^2(t)]$ given by Eq. (4.36) via the calculus of variations; we shall use the orthogonality principle given by Theorem 1. We can now state the solution for the optimal filter by the following theorem.

Theorem 3 (Wiener-Hopf)

$E[e_\alpha^2(t)]$ given by Eq. (4.29) is minimized if and only if $\hat{h}(t)$ can be obtained from the solution of the equation:

$$R_s(\tau + \alpha) = R_{sY}(\tau + \alpha) = \int_{-\infty}^{\infty} \hat{h}(\sigma)\, R_Y(\tau - \sigma)\, d\sigma$$

$$= \int_{-\infty}^{\infty} \hat{h}(\sigma)\, [R_s(\tau - \sigma) + R_n(\tau - \sigma)]\, d\sigma$$

$$(4.37)$$

Thus, the optimal solution $\hat{s}(t)$ is given via:

$$\hat{s}(t) = \int_{-\infty}^{\infty} Y(\lambda)\, \hat{h}(t-\lambda)\, d\lambda = \int_{-\infty}^{\infty} \hat{h}(\lambda)\, Y(t-\lambda)\, d\lambda \qquad (4.38)$$

Equation (4.37) is known as the Wiener-Hopf equation.

Proof

We have proven the orthogonality principle for the discrete case. In what follows we shall show that the solution of Eq. (4.37) is equivalent to the solution of:

$$E[e_\alpha(t)\, Y(\theta)] = E[(s(t+\alpha) - \hat{s}(t))\, Y(\theta)] = 0, \text{ for all } \theta \leqslant t \qquad (4.39)$$

where $\hat{s}(t)$ is given by Eq. (4.38), $\theta = t - \tau$ with $0 < \tau < \infty$. Let us use the notation $\hat{s}(t_1 | t)$ as the optimal estimate to Eq. (4.29), given the observation $Y(t)$ over $(-\infty, t]$, where $t_1 = t + \alpha$.

To prove (4.39), let $\mathscr{V}$ be the space generated by the random variable $\{s(t_1)\}$. Let $Q \subset \mathscr{V}$ be a space generated from $\{Y(t)\}$ given by elements:

$$q(t_1) = \int_{-\infty}^{\infty} h(t_1 - \tau)\, Y(\tau)\, d\tau$$

where $h(\cdot)$ is a continuously differentiable function. Utilizing Theorem 1, the norm (mean square)

$$\|s - q\|_{q.m.}$$

is minimized if $\hat{q} = Ps \in Q$ and from Claim 2:

$$E[(\hat{s}(t_1 | t) - s(t))\, q(t_1)] = 0$$

which yields:

$$E[e(t_1 | t)\, q(t_1)] = \int_{-\infty}^{\infty} R_{eY}(t_1 - \tau | t)\, h(t_1 - \tau)\, d\tau = 0$$

which proves the orthogonality condition.

4.7.1 Discussion

The Wiener-Hopf equation (4.37) will provide the solution for $\hat{h}(t)$. However, obtaining $\hat{h}(t)$ from the integral equation is extremely difficult. Assuming the observation $Y(t)$ is available over the interval $(-\infty, t)$, we can utilize the frequency domain approach to solve for $\hat{h}(t)$ by obtaining $\hat{H}(j\omega)$.

It turns out that $\hat{h}(t)$ does not correspond to a causal system (realizable), since, in general, $\hat{h}(t)$ is non-zero for $t < 0$. The condition of realizability is given by the Paley-Wiener condition (a sufficiency condition) which states that a system with the transfer function $H(j\omega) = \mathscr{F} h(t)$ is realizable only if

$$\int_{-\infty}^{\infty} \frac{|\ln|H(j\omega)||}{1+\omega^2} \, d\omega < \infty \tag{4.40}$$

The linear system described above will, in general, violate condition (4.40).

If we drop the condition of realizability for the moment, we obtain (to be proven) $\hat{H}(j\omega)$ as:

$$\hat{H}(j\omega) = \frac{S_{sY}(\omega) \exp(j\omega\alpha)}{S_Y(\omega)} = \frac{S_s(\omega) \exp(j\omega\alpha)}{S_Y(\omega)} \tag{4.41}$$

Hence, $\hat{h}(t)$ can be obtained as the inverse Fourier transform of $\hat{H}(j\omega)$. Thus,

$$\hat{h}(t) = \frac{1}{2\pi} \int_{-\infty}^{\infty} \frac{S_s(\omega) \exp[j\omega(t+\alpha)]}{S_Y(\omega)} d\omega \tag{4.42}$$

Let C denote $E[e_\alpha^2(t)]$ and C^0 its minimum over all $h(\cdot)$. We shall also prove that:

$$C^0 = \frac{1}{2\pi} \int_{-\infty}^{\infty} \frac{S_s(\omega) S_n(\omega)}{S_Y(\omega)} d\omega \tag{4.43}$$

Remark 2. If $s(t)$ and $n(t)$ are uncorrelated, then

$$R_{sY}(\tau) = R_s(\tau) \leftrightarrow S_{sY}(\omega) = S_Y(\omega) \tag{4.44}$$

Remark 3. Utilizing the orthogonality principle (see Eq. 4.39), we can verify that:

$$C^0 = E\{[s(t) - \hat{s}(t)]^2\} = E[s^2(t)] - E[\hat{s}^2(t)]$$

$$= E\{[s(t) - \hat{s}(t)]\, s(t)\}$$

Thus,

$$C^0 = R_s(0) - \int_{-\infty}^{\infty} R_{sY}(\tau)\, h(\tau)\, d\tau \qquad (4.45)$$

Theorem 4

The optimal transfer function $\hat{H}(j\omega)$ corresponding to the impulse response is given by Eq. (4.41) and C^0 given by Eq. (4.45) is the minimum (optimal) performance index.

Proof

From the Wiener-Hopf equation, we have:

$$R_{sY}(\tau + \alpha) = R_s(\tau + \alpha) = \int_{-\infty}^{\infty} \hat{h}(\sigma)\, R_Y(\tau - \sigma)\, d\sigma$$

Now let us take the Fourier transform of the above:

$$\exp(j\omega\alpha)\, S_{sY}(\omega) = \int_{-\infty}^{\infty} R_{sY}(\tau + \alpha)\, \exp(-j\omega\tau)\, d\tau$$

$$= \int_{-\infty}^{\infty} \int_{-\infty}^{\infty} \hat{h}(\sigma)\, R_Y(\tau - \sigma)\, \exp(-j\omega\tau)\, d\sigma\, d\tau$$

$$= \underbrace{\int_{-\infty}^{\infty} \hat{h}(\theta)\, \exp(-j\omega\theta)\, d\theta}_{\hat{H}(j\omega)} \underbrace{\int_{-\infty}^{\infty} R_Y(\lambda)\, \exp(-j\omega\lambda)\, d\lambda}_{S_Y(\omega)}$$

Thus,

$$\hat{H}(j\omega) = \frac{S_{sY}(\omega)\exp(j\omega\alpha)}{S_Y(\omega)} = \frac{S_s(\omega)\exp(j\omega\alpha)}{S_Y(\omega)}$$

as asserted.

To prove Eq. (4.43), let us calculate $C = E[e_\alpha^2(t)]$ via the frequency domain. From Eq. (4.36), we know

$$C = R_s(0) - 2\int h(\tau) R_s(\tau + \alpha)\, d\tau$$

$$+ \int\int h(\tau) h(\sigma) R_Y(\tau - \sigma)\, d\tau\, d\sigma$$

Thus, C can be rewritten as:

$$C = \frac{1}{2\pi}\int S_s(\omega)\, d\omega - \frac{2}{2\pi}\int h(\tau)\int S_s(\omega) \exp[j\omega(\tau + \alpha)]\, d\omega\, d\tau$$

$$+ \frac{1}{2\pi}\int\int h(\tau) h(\sigma)\int S_Y(\omega) \exp[-j\omega(\sigma - \tau)]\, d\omega\, d\sigma\, d\tau$$

$$= \frac{1}{2\pi}\int S_s(\omega)\, d\omega - \frac{2}{2\pi}\int h(\tau)\exp(j\omega\tau)\, d\tau \int S_s(\omega)\exp(j\omega\alpha)\, d\omega$$

$$+ \frac{1}{2\pi}\int h(\tau)\exp(j\omega\tau)\, d\tau \int h(\sigma)\exp(-j\omega\sigma)\, d\sigma \int S_Y(\omega)\, d\omega$$

$$= \frac{1}{2\pi}\int \left\{ S_s(\omega)[|1 - 2H^*(j\omega)\exp(j\omega\alpha)|] + |H(j\omega)|^2 S_Y(\omega) \right\} d\omega$$

Thus,

$$C = \frac{1}{2\pi} \int \left\{ S_s(\omega)[|\exp(j\omega\alpha) - H(j\omega)|^2] + |H(j\omega)|^2 S_n(\omega) \right\} d\omega$$

Now if we substitute $\hat{H}(j\omega)$ from Eq. (4.41) into the above equation, we obtain:

$$C^0 = \frac{1}{2\pi} \int \left[S_s(\omega) \left| \exp(j\omega\alpha) - \frac{\exp(j\omega\alpha) S_s(\omega)}{S_Y(\omega)} \right|^2 + \frac{S_s^2(\omega) S_n(\omega)}{S_Y^2(\omega)} \right] d\omega$$

$$= \frac{1}{2\pi} \int \frac{S_s(\omega) S_n(\omega)}{S_Y(\omega)} d\omega$$

The proof is now complete.

Example 3

Assume that the signal $s(t)$ and the noise $n(t)$ are uncorrelated and that they are both of zero mean. Let

$$S_s(\omega) = \frac{1}{1 + \omega^2}$$

and

$$S_N(\omega) = 1$$

Obtain the optimal estimate $\hat{s}(t)$ of $s(t + \alpha)$.

Solution

Since the noise and the signals are uncorrelated, then $S_Y(\omega) = S_s(\omega) + S_n(\omega)$. If $\alpha = 0$, then

$$\hat{H}(j\omega) = \frac{\frac{1}{1 + \omega^2}}{\frac{1}{1 + \omega^2} + 1} = \frac{1}{2 + \omega^2}$$

and

$$\hat{h}(t) = \frac{1}{2\sqrt{2}} \exp[-\sqrt{2}\,|t|]$$

For prediction and smoothing $\alpha \neq 0$, then

$$\hat{H}(j\omega) = \frac{1}{2 + \omega^2} \exp(j\omega\alpha)$$

and $\hat{h}(t)$ is its inverse Fourier transform.

4.7.2 A Very Important Remark

Although the optimal impulse response $\hat{h}(t)$ corresponding to $\hat{s}(t)$ is not realizable, it can be solved mathematically. We have solved for $\hat{H}(j\omega)$ by utilizing the frequency domain analysis, where $\hat{h}(t)$ is the inverse Fourier transform of $\hat{H}(j\omega)$. We should emphasize that the solution was possible in closed form (see Eqs. 4.41–4.43) by making some significant assumptions:

(a) First, we assumed that the measurement $Y(t)$ passes through a time-invariant linear system (filter).

(b) The measurement of the observation $Y(t)$ was available over the semi-infinite interval.

Assumptions (a) and (b) were made so that we could utilize the frequency domain approach to solve the complicated Wiener-Hopf equation.

4.7.3 Wiener-Kolmogorov Theory for the Time-Varying Case

It should be emphasized that the Wiener-Kolmogorov theory does not have to satisfy assumptions (a) and (b). In that case, the optimal linear system will be time-varying, and we would not be able to use the frequency domain analysis to advantage.

The W-K theory for the time-varying case assumes the availability of the observation $Y(t)$ over the finite interval $[t_0, t]$. Now we will seek a time-varying impulse function $\hat{h}(t, \tau)$ such that (see sketch):

$$\hat{s}(t) = \int_{t_0}^{t} \hat{h}(t, \tau)\, Y(\tau)\, d\tau \qquad (4.46)$$

where

$$C^0 = \min E[e_\alpha^2(t)] = E\{[s(t+\alpha) - \hat{s}(t)]^2\} \qquad (4.47)$$

over all $h(t, \tau)$.

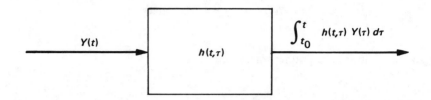

We now state a general theorem concerning the optimal solution.

Theorem 5 (Wiener-Hopf)

The optimal solution $\hat{s}(t)$ given by Eq. (4.46) is obtained if and only if $\hat{h}(t, \tau)$ is solved from:

$$R_{sY}(t - \alpha) = \int_{t_0}^{t} \hat{h}(t, \sigma) R_Y(\sigma - \alpha) \, d\sigma \qquad (4.48)$$

Proof

The proof of the Wiener-Hopf equation given by (4.46) is equivalent to the orthogonality principle:

$$E[e_\alpha(t) Y(\theta)] = 0, \quad t_0 \leq \theta \leq t$$

as already discussed by Theorem 3. The proof is identical to that of Theorem 3 with the only difference being that the integral limits are from t_0 to t and $h(t - \tau)$ is replaced by $h(t, \tau)$.

Note that if the power spectrums of $n(t)$ and $s(t)$ do not overlap (see sketch), then $S_s(\omega) S_n(\omega) = 0$ and from Eq. (4.43), we get:

$$C^0 = 0$$

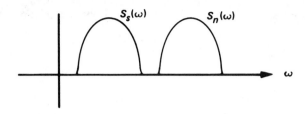

Thus, there is no error in the system. Hence, we can separate the signal and the noise perfectly.

4.8 OPTIMUM CAUSAL SYSTEMS

Now we shall seek an optimum system which is constrained to be physically realizable, i.e., the impulse response should be $\hat{h}(\lambda) = 0$ whenever $\lambda < 0$. Thus, from Eq. (4.38):

$$\hat{s}(t) = \int_0^\infty \hat{h}(\lambda)\, Y(t - \lambda)\, d\lambda \tag{4.49}$$

that is, $\hat{s}(t)$ is not a function of $Y(t - \lambda)$ for $\lambda < 0$, which is not available, since $\hat{h}(\lambda) = 0$ for $\lambda < 0$. The upper bound is ∞, since the observation over the interval $[-\infty, t]$ is available to the estimator.

Without any loss of generality assume that $\alpha = 0$. Then the orthogonality principle is:

$$E\{[s(t) - \hat{s}(t)]\, Y(t - \tau)\} = 0, \quad \text{for } 0 \leq \tau < \infty \tag{4.50}$$

and its corresponding Wiener-Hopf equation is:

$$R_{sY}(\tau) = \int_0^\infty \hat{h}(\sigma)\, R_Y(\tau - \sigma)\, d\sigma \tag{4.51}$$

or

$$R_{sY}(\tau) = R_{\hat{s}Y}(\tau), \quad \text{for all } \tau > 0 \tag{4.52}$$

(see Eq. 4.50).

Let $q(\tau)$ be defined by:

$$q(\tau) = R_{sY}(\tau) - R_{\hat{s}Y}(\tau), \quad \text{for all } \tau \tag{4.53}$$

Note that for all $\tau > 0$, $q(\tau) = 0$. Taking the Fourier transform of the above yields:

$$Q(\omega) = S_{sY}(\omega) - S_{\hat{s}Y}(\omega) = S_{sY}(\omega) - \hat{H}(j\omega) S_Y(\omega) \qquad (4.54)$$

assuming $Q(\omega)$ exists. Now replace $\omega = s/j$ in Eq. (4.54) to get the bilateral Laplace transform:

$$Q(s) = S_{sY}(s) - S_{\hat{s}Y}(s) = S_{sY}(s) - \hat{H}(s) S_Y(s) \qquad (4.55)$$

We have already discussed the fact that the bilateral transform $F(s)$ of any absolutely integrable function $f(t)$, for $t > 0$, will have poles in the left-half plane (LHP), and, for $t < 0$, will have poles in the right-half plane (RHP). Thus, $Q(s)$ cannot have LHP poles since $q(\tau) = 0$ for all $\tau > 0$. We know $S_Y(s)$ is an even function of s; let us decompose it as follows:

$$S_Y(s) = S_Y^+(s) S_Y^-(s) \qquad (4.56)$$

Where $S_Y^+(s)$ will have LHP poles and $S_Y^-(s)$ will have RHP poles (that is, $S_Y^+(s)$ is analytic in the RHP and $S_Y^-(s)$ is analytic in the LHP). Using Eq. (4.56) in Eq. (4.55) yields:

$$Q(s) = S_{sY}(s) - \hat{H}(s) S_Y^+(s) S_Y^-(s) \qquad (4.57)$$

From Eq. (4.57) we obtain:

$$\hat{H}(s) S_Y^+(s) = \frac{S_{sY}(s)}{S_Y^-(s)} - \frac{Q(s)}{S_Y^-(s)} \qquad (4.58)$$

We observe that $\hat{H}(s) S_Y^+(s)$ has its poles in the LHP and $Q(s)/S_Y^-(s)$ has all its poles in the RHP. But $S_{sY}(s)/S_Y^-(s)$ has poles all over the complex plane.

Let

$$G(s) = \frac{S_{sY}(s)}{S_Y^-(s)} \qquad (4.59)$$

The partial fraction expansion of $G(s)$ can be decomposed as:

$$G(s) = G_1(s) + G_2(s) \qquad (4.60)$$

where $G_1(s)$ will have LHP poles and $G_2(s)$ will have RHP poles only.

Now choose (see Eq. 4.58):

$$\widehat{H}(s) = \frac{G_1(s)}{S_Y^+(s)} \qquad (4.61)$$

Thus, $\widehat{H}(s)$ given by Eq. (4.61) is the solution of the Wiener-Hopf equation. The above solution is due to Shannon and Bode.

The following examples are taken from reference [8].

Example 4

Let $s(t)$ and $n(t)$ be stochastic signals of zero mean, such that:

$$R_s(\tau) = \frac{3}{2}\exp(-|\tau|)$$

$$R_n(\tau) = \delta(\tau)$$

and

$$E[s(t)\,n(t)] = 0$$

Let us derive an optimal $\hat{s}(t)$ of $s(t)$ over $(-\infty, t)$.

Solution

From Eq. (4.58):

$$\widehat{H}(s)\, S_Y^+(s) = \frac{S_{sY}(s)}{S_Y^-(s)} - \frac{Q(s)}{S^-(s)}$$

and from (4.61):

$$\hat{H}(s) = \frac{G_1(s)}{S_Y^+(s)}$$

where $G_1(s)$ will correspond to the LHP poles of $S_{sY}(s)/S_Y^-(s)$, upon partial fraction expansion, we get:

$$S_{sY}(\omega) = S_s(\omega) = \frac{3}{1+\omega^2}$$

$$S_Y(\omega) = 1 + S_s(\omega) = \frac{4+\omega^2}{1+\omega^2}$$

The bilateral Laplace transform corresponding to $S_Y(\omega)$ is:

$$S_Y(s) = \frac{(2+s)(2-s)}{(1+s)(1-s)}$$

or

$$S_Y^+(s) = \frac{2+s}{1+s}$$

and

$$\frac{S_s(s)}{S_Y^-(s)} = \frac{3/(1-s^2)}{(2-s)/(1-s)} = \frac{3}{(1+s)(2-s)} = \frac{1}{1+s} + \frac{1}{2-s}$$

Hence,

$$G_1(s) = \frac{1}{1+s} \text{ and } \hat{H}(s) = \frac{1}{2+s} \leftrightarrow \hat{h}(t) = \exp(-2t)$$

See sketch below.

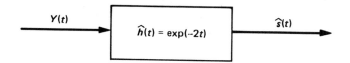

Sometimes we designate the procedure by the block diagram:

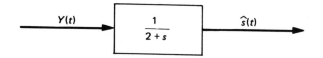

The filter can also be written as a differential equation:

$$\dot{\hat{s}}(t) = -2\hat{s}(t) + Y(t)$$

Example 5

Let $s(t)$ and $n(t)$ be given such that:

$$R_s(\tau) = \exp(-|\tau|)$$

$$R_{sn}(\tau) = 0$$

$$R_n(\tau) = \exp(-2|\tau|)$$

where $s(t)$ and $n(t)$ are of zero mean. Let $Y(\lambda) = s(\lambda) + n(\lambda)$ be given on $(-\infty, t]$. Find the optimal estimate $\hat{s}(t)$ of $s(t)$.

Solution

$$S_{sY}(\omega) = S_s(\omega) = \frac{2}{1 + \omega^2}$$

and

$$S_y(\omega) = S_s(\omega) + S_n(\omega) = \frac{6(2 + \omega^2)}{(1 + \omega^2)(4 + \omega^2)}$$

149

Now the bilateral transform corresponding to $S_Y(\omega)$ is obtained by inspection, and $S_Y^+(s)$ is given by:

$$S_Y^+(s) = \frac{\sqrt{6}\,(\sqrt{2}+s)}{(1+s)(2+s)}$$

Also, the partial fraction expansion of $S_s(s)/S_Y^-(-s)$ must be obtained:

$$S_s(s)/S_Y^-(s) = \frac{2/(1-s^2)}{\sqrt{6}\,(\sqrt{2}-s)/(1-s)(2-s)} = G_1(s) + G_2(s)$$

where

$$G_1(s) = \frac{\sqrt{6}}{1+\sqrt{2}}\left(\frac{1}{1+s}\right)$$

after partial fraction expansion. Thus,

$$\widehat{H}(s) = \left(\frac{1}{1+\sqrt{2}}\right)\left(\frac{2+s}{\sqrt{2}+s}\right)$$

The optimum filter is given via the figure.

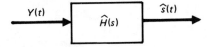

4.8.1 Optimal Prediction and Smoothing

We have thus far obtained the optimal estimate $\hat{s}(t)$ of $s(t)$ given $Y(t)$ on the interval $(-\infty, t]$, i.e., we have derived the optimal filter. Remember that $\hat{s}(t)$ is the output of the linear system with the impulse response $\hat{h}(t)$ and the input $Y(t)$. Suppose we are interested in estimating $s(t+t_0)$ based on the same observation $Y(t)$ on $(-\infty, t]$, where $t_0 > 0$. This is called prediction. Before obtaining the optimal predictor $\hat{s}(\cdot)$, let us generalize the estimation problem somewhat.

Let $s(t)$ and $n(t)$ be as before, i.e., they are zero mean and wide-sense stationary such that:

$$R_{sn}(\tau) = 0$$

Define

$$W(t) = \int_{-\infty}^{\infty} g(\lambda)\, s(t - \lambda)\, d\lambda \qquad (4.62)$$

where $g(\cdot)$ is a fixed impulse response of a time-invariant system.

Now given the observation $Y(t)$, let us minimize

$$C = E[W(t) - W_0(t)]^2 \qquad (4.63)$$

where $W_0(t)$ is the output of the filter with impulse response $h(t)$ and input $Y(t)$, and $h(\cdot)$ is restricted to be causal. The optimal solution $\widehat{W}(t)$ is thus the output of the system with impulse response $\hat{h}(t)$ and input $Y(t)$.

Equations (4.62) and (4.63) are the generalization of the filtering problem. For example, if $g(t) = \delta(t)$, then $W(t) = s(t)$. Thus, in Eq. (4.63) we are minimizing the difference between a linear function of the signal given by Eq. (4.62) and a linear function of the observation. The impulse response $g(t)$ is fixed in each case; however, we must determine an optimal $\hat{h}(t)$ from the class of all impulse responses $h(t)$, such that Eq. (4.63) is minimized. For example, for the case of prediction and smoothing, where $W(t) = s(t \pm t_0)$ with $t_0 > 0$, $W(t)$ can be considered to be the output of a linear time-invariant system with input $s(t)$ and impulse response $g(t) = \delta(t \pm t_0)$.

If it is desired to derive an estimator for the system

$$W(t) = s(t) + s(t + kt_0) \qquad (4.64)$$

where k and t_0 are positive numbers, then the impulse response of the corresponding filter is:

$$g(t) = [\delta(t) + \delta(t + kt_0)].$$

If $W(t)$ is given by:

$$W(t) = \int_0^{\infty} \exp(-a\lambda)\, s(t - \lambda)\, d\lambda = \int_{-\infty}^{t} \exp[-a(t - \lambda)]\, s(\lambda)\, d\lambda$$

then the impulse response $g(t)$ is $\exp(-at)$, $t \geq 0$.

Let $\mathcal{G}(j\omega)$ denote the Fourier transform of $g(t)$. Then following the same procedure as from Eq. (4.50) to (4.61) we obtain:

$$\frac{\mathcal{G}(s) S_{sY}(s)}{S_Y^-(s)} = G_1(s) + G_2(s) \qquad (4.65)$$

where (4.65) is a generalization of (4.60).

Now $\hat{H}(s)$ is given by:

$$\hat{H}(s) = \frac{G_1(s)}{S_Y^+(s)} \qquad (4.66)$$

Remark 4. In Examples 4 and 5, $\mathcal{G}(s) = 1$.

Example 6

Use Example 4 to obtain the best estimate of $s(t + t_0)$, $t_0 > 0$.

Solution

$g(\lambda) = \delta(\lambda + t_0)$ or $\mathcal{G}(s) = \exp(t_0 s)$. Thus, as before,

$$S_Y^+(s) = \frac{2+s}{1+s}$$

Now, due to the factor $\exp(t_0 s)$, the decomposition of $\mathcal{G}(s) S_{sY}(s)/S_Y^-(s)$ is given by:

$$\frac{\mathcal{G}(s) S_{sY}(s)}{S_Y^-(s)} = G_1(s) + G_2(s)$$

However, let us derive the portion of the function $\mathcal{G}(s) S_{sY}(s)/S_Y^-(s)$ corresponding to $t > 0$ or $G_1(s)$. Thus,

$$G_1(s) = \frac{\exp(-t_0)}{1+s}$$

therefore, $\widehat{H}(s)$ is then given by:

$$\widehat{H}(s) = \frac{\exp(-t_0)}{2+s}$$

For smoothing, the results are similar.

Example 7

In Example 4, obtain the best estimator for $W(t)$ given by Eq. (4.64) with $k = 2$.

Solution

$$W(t) = \int_{-\infty}^{\infty} g(t - \lambda) s(\lambda) \, d\lambda$$

where $g(t)$ is given by:

$$g(t) = [\delta(t) + \delta(t + 2t_0)]$$

Thus,

$$G(s) = \frac{3[1 + \exp(2t_0 s)]}{(1 + s)(2 - s)}$$

where

$$G_1(s) = \frac{[1 + \exp(-2t_0)]}{1 + s}.$$

Hence,

$$\widehat{H}(s) = \frac{G_1(s)}{S_Y^+(s)} = \frac{[1 + \exp(-2t_0)]}{2 + s}$$

153

4.9 MATCHED FILTERING

In laser and radar applications, when a system is used to detect a target, the form of the signal must be known. However, often the signal is contaminated by additive noise. A good criterion for estimation could be the signal-to-noise ratio (SNR), which we would be interested in maximizing.

Now let us assume that $s(t)$ is a deterministic signal such that its Fourier transform (denoted by $S(\omega)$) exists. Let $S_n(\omega)$ be the power spectrum of the noise contaminating the signal. Let both the signal and the noise pass through a time-invariant system with the transfer function $H(j\omega)$, and let $Y_s(t)$ denote the output corresponding to $s(t)$ with $Y_n(t)$ the output corresponding to $n(t)$.

Suppose at $t = t_1$, we are interested in maximizing

$$\rho = \frac{Y_s^2(t_1)}{E(Y_n^2(t_1))} \tag{4.67}$$

$Y_s^2(t)$ is the output power of the signal, and we know that $E(Y_n^2(t))$ is the output power due to noise. We can write Eq. (4.67) in terms of the frequency parameter. We know that:

$$Y_s(t) = \int_{-\infty}^{\infty} h(t-\tau)\, s(\tau)\, d\tau \tag{4.68}$$

and

$$Y_n(t) = \int_{-\infty}^{\infty} h(t-\tau)\, n(\tau)\, d\tau \tag{4.69}$$

Also note that:

$$S_{Y_n}(\omega) = |H(j\omega)|^2\, S_n(\omega) \tag{4.70}$$

and

$$\mathscr{F}\{Y_s(t)\} = H(j\omega)\, S(\omega) \tag{4.71}$$

Thus, from (4.71), $Y_s(t)$ can be obtained as $\mathscr{F}^{-1}$ of $H(j\omega) S(\omega)$, i.e.,

$$Y_s(t) = \frac{1}{2\pi} \int_{-\infty}^{\infty} H(j\omega) S(\omega) \exp(j\omega t) \, d\omega \qquad (4.72)$$

and $E[Y_n^2(t)]$ as the $\mathscr{F}^{-1}$ of $S_{Y_n}(\omega)$. Thus,

$$E[Y_n^2(0)] = E[Y_n^2(t)] = \frac{1}{2\pi} \int_{-\infty}^{\infty} |H(j\omega)|^2 S_n(\omega) \, d\omega \qquad (4.73)$$

If we are interested in maximizing the SNR given (4.67) at $t = t_1$, we must maximize:

$$\rho = \frac{Y_s^2(t_1)}{E[Y_n^2(t_1)]} = \frac{\left[\int_{-\infty}^{\infty} H(j\omega) S(\omega) \exp(j\omega t_1) \, d\omega\right]^2}{2\pi \int_{-\infty}^{\infty} |H(j\omega)|^2 S_n(\omega) \, d\omega} \qquad (4.74)$$

We now state and prove the following theorem.

Theorem 6

The maximum value of the signal-to-noise ratio ρ given by Eq. (4.74) is obtained if:

$$H(j\omega) = k \frac{S^*(\omega)}{S_n(\omega)} \exp(-j\omega t_1) \qquad (4.75)$$

where k is a constant. Before proving the above, we note the following:

The intuitive concept of Eq. (4.75) is obvious: The filter should pass those frequencies for which the amplitude spectrum of the signal is large compared to $S_n(\omega)$, which is the power spectrum of the noise.

The special case where $S_n(\omega)$ is constant, say, $\mathcal{N}_0$, is very important, i.e., white noise. In that case Eq. (4.75) becomes:

$$H(j\omega) = \frac{k}{\mathcal{N}_0} S^*(\omega) \exp(-j\omega t_1) \tag{4.76}$$

The factor $k/\mathcal{N}_0$ is gain, which we shall assume is unity without any loss of generality. Since the transfer function that maximizes ρ is given by the conjugate of $S(\omega)$ (and $\exp(-j\omega t_1)$), the filter $H(j\omega)$ is called the conjugate filter. However, a more popular definition is the match filter, since $H(j\omega)$ is to match $S^*(\omega) \exp(-j\omega t_1)$.

Proof of Theorem 6

The proof is relatively simple. Using the Cauchy-Schwarz inequality:

$$\left| \int f(\omega) g(\omega) d\omega \right|^2 \leq \int |f(\omega)|^2 d\omega \int |g(\omega)|^2 d\omega \tag{4.77}$$

we set:

$$f(\omega) = H(j\omega) [S_n(\omega)]^{1/2}$$

and

$$g(\omega) = \frac{S(\omega) \exp(j\omega t_1)}{[S_n(\omega)]^{1/2}}$$

The left-hand-side, when divided by the first integral on the right, is simply $2\pi\rho$, which implies:

$$\rho \leq \frac{1}{2\pi} \int_{-\infty}^{\infty} \frac{|S(\omega)|^2}{S_n(\omega)} d\omega \tag{4.78}$$

As a consequence of the Cauchy-Schwarz inequality, if $f(\omega) = k g^*(\omega)$, then we shall have the equality in (4.77). Therefore, ρ becomes maximum if:

$$H(j\omega) = k \frac{S^*(\omega)}{S_n(\omega)} \exp(-j\omega t_1)$$

Thus, the proof is completed.

4.10 KALMAN-BUCY FILTERING

Before discussing Kalman filtering, let us review some basic concepts needed in the discussion.

Definition 3

A continuous Markov process $X(t)$ for $t > t_0$ is a process that, for every $\tau \leq t$,

$$f(X(t)|X(\lambda), \text{ for } \lambda \in [t_0, \tau]) = f(X(t)|X(\tau)) \tag{4.79}$$

where λ can assume any value in the interval $t_0 \leq \lambda \leq \tau \leq t$. For the discrete case the definition is similar. Let $t_0, t_1, t_2, \ldots, t_n$ be such that:

$$t_0 < t_1 < t_2 < \ldots < t_n \tag{4.80}$$

and $\{X(\cdot)\}$ be a discrete set of random variables taking on the values from $\{t_i\}_{i=1}^{n}$. Let us use the notation $X(i)$ instead of $X(t_i)$. We can now define the discrete Markov process.

Definition 4

The process $\{X(i)\}$ is a Markov process if for every n such that (4.80) is satisfied, we have:

$$f(X(n)|X(0), X(1), \ldots, X(n-1)) = f(X(n)|X(n-1)) \tag{4.81}$$

Now utilizing:

$$f(X(0), X(1), \ldots, X(n)) = f(X(0), X(1), \ldots, X(n-1))f(X(n)|X(0),$$
$$X(1), \ldots, X(n-1))$$

and continuing in this manner, and making use of definition (4.81), we get:

$$f(X(0), X(1), \ldots, X(n)) = f(X(0))f(X(1)|X(0)) \cdots f(X(n)|X(n-1))$$

$$= f(X(0)) \prod_{i=1}^{n} f(X(i)|X(i-1)) \tag{4.82}$$

Hence, the Markov process is defined by the conditional probability density functions $f(X(i)|X(i-1))$ for $i = 1, \ldots, n$. The Markov process is fundamental to Kalman-Bucy filter development.

As already discussed, a linear system can be characterized via the classical method using the impulse response or the modern approach using the state variable approach. Kalman-Bucy filtering relies on the state variable characterization, where the state is a Markov process.

The reader is assumed to be familiar with the simple state variable representation. If this familiarity does not exist, the reader should consult Appendix E, which contains a simplified discussion of state variables along with some examples. That appendix is sufficient for our purposes.

4.10.1 Continuous Kalman-Bucy Recursive Filtering

We shall briefly discuss the continuous version of Kalman-Bucy (K-B) filtering. The most important part of K-B filtering is the fact that estimation is of a sequential nature (Markovian). We shall discuss K-B filtering for linear systems unless specified otherwise.

The state variable characterization of a linear system can be generally written as:

$$\dot{X} = A(t) X(t) + B(t) U(t) \quad \text{(a)}$$
$$Y(t) = C(t) X(t) + D(t) U(t) \quad \text{(b)}$$
(4.83)

where $X(t) = [X_1(t), \ldots, X_n(t)]'$, where the prime denotes the transpose, $U(t)$ is a $p \times 1$ matrix, and $Y(t)$ is a $q \times 1$ matrix. $A(t), B(t), C(t), D(t)$ are matrices of order $n \times n, n \times p, q \times n,$ and $q \times p$, respectively.

Example 8

Let a time-invariant system be characterized by the following differential equation:

$$\frac{d^3 Y(t)}{dt^3} + 2 \frac{d^2 Y(t)}{dt^2} + 3 \frac{dY(t)}{dt} + Y(t) = 2U(t) \qquad (4.84)$$

where $Y(t)$ is the output, $U(t)$ the input.

Define the state variables as follows:

$$X_1(t) = Y(t) \tag{4.85}$$

$$X_2(t) = \frac{dX_1(t)}{dt} = \frac{dY(t)}{dt} \tag{4.86}$$

$$X_3(t) = \frac{dX_2(t)}{dt} = \frac{d^2 Y(t)}{dt^2} \tag{4.87}$$

Equation (4.84) can be arranged so that the highest-order derivative term appears on one side of the equation. Thus,

$$\frac{d^3 Y(t)}{dt^2} = -2\frac{d^2 Y(t)}{dt^2} - 3\frac{dY(t)}{dt} - Y(t) + 2U(t) \tag{4.88}$$

Substituting (4.85) – (4.87) into (4.88) and utilizing the defining relations of the state variable into (4.88) yields:

$$\dot{X}_1 = X_2(t) \tag{4.89a}$$

$$\dot{X}_2 = X_3(t) \tag{4.89b}$$

$$\dot{X}_3 = -X_1(t) - 3X_2(t) - 2X_3(t) + 2U(t) \tag{4.89c}$$

The system described by (4.84) can then be defined by the state variable representation of the form (4.83). Thus,

$$A = \begin{bmatrix} 0 & 1 & 0 \\ 0 & 0 & 1 \\ -1 & -3 & -2 \end{bmatrix}$$

$$B = \begin{bmatrix} 0 \\ 0 \\ 2 \end{bmatrix}$$

$$C = [1 \quad 0 \quad 0]$$

$$D = 0$$

The solution of $X(t)$ is given by:

$$X(t) = \Phi(t, t_0) X(t_0) + \int_{t_0}^{t} \Phi(t, \tau) B(\tau) U(\tau) \, d\tau \qquad (4.90)$$

where

$$\frac{d\Phi(t, t_0)}{dt} = A(t) \Phi(t, t_0) \qquad (4.91)$$

$$\Phi(t_0, t_0) = I \text{ (identity matrix)} \qquad (4.92)$$

$\Phi(t, t_0)$ is called the transition matrix, which is a matrix of order $n \times n$. Furthermore, it can be shown that (see Appendix E) the following relations hold:

$$\Phi^{-1}(t_1, t_0) = \Phi(t_0, t_1) \qquad (4.93)$$

$$\Phi(t_2, t_0) = \Phi(t_2, t_1) \Phi(t_1, t_0) \qquad (4.94)$$

and Φ is a nonsingular matrix.

In a time-invariant system (A, B, C, and D are constant matrices), the transition matrix $\Phi(t, t_0)$ takes the form:

$$\Phi(t, t_0) = \exp \{A \cdot (t - t_0)\}$$

where

$$\exp \{A \cdot t\} \triangleq I + At + \frac{A^2 t^2}{2!} + \ldots + \frac{A^n t^n}{n!} + \ldots$$

A general diagram of the system given by Eq. (4.83) is given in Figure 4-3.

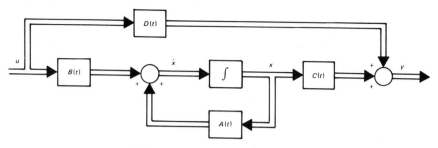

Fig. 4-3. State Variable Configuration

The continuous Kalman estimation requires a linear system model of the form:

$$\dot{X} = A(t) X(t) + B(t) U(t) \qquad (4.95)$$

$$Y(t) = C(t) X(t) + v(t) \qquad (4.96)$$

where $X(t)$ is assumed to be a random process, an $n \times 1$ matrix, $U(t)$ a random noise of zero mean, a $p \times 1$ matrix, $v(t)$ is a random noise with zero mean and a $q \times 1$ matrix uncorrelated with $U(t)$. $A(t)$, $B(t)$, and $C(t)$ are matrices of dimensions $n \times n$, $n \times p$, $q \times n$, respectively. The observation signal $Y(t)$ is contaminated by the additive noise process $v(t)$. The most important property of Kalman estimation is the fact that a differential equation technique developed to solve the optimal solution has the property that it can be synthesized in a recursive manner because the differential equation techniques are in most instances equivalent or very closely related to recursive techniques. That is, the estimate at one point does not need the processing of all the measurements, but only the information stored by the point preceding it.

Let us assume the following statistical moments:

$$EU(t) = 0$$

$$Ev(t) = 0$$

$$EU(t_1) U'(t_2) = Q\delta(t_2 - t_1) \qquad (4.97)$$

$$Ev_1(t_1) v'(t_2) = L\delta(t_2 - t_1)$$

$$EU(t_1) v'(t_2) = 0$$

where Q and L are of dimensions $p \times p$ and $q \times q$, respectively. These matrices are generally functions of time t, and $\delta(t_2 - t_1)$ is the Dirac delta function. The functions $U(t)$ and $v(t)$ are white noise terms with respective covariances Q and L.

The Kalman recursive problem is one in which we are given the observation values (continuous measurements) of $Y(\tau)$, $t_0 \leq \tau \leq t$, and it is desired to find the estimate at time t_1 denoted as $\hat{X}(t_1|t)$ or $X(t_1)$ having the form:

$$\hat{X}(t_1|t) = \int_{t_0}^{t} h(t, \tau) Y(\tau) d\tau$$

where $h(t, \tau)$ is the impulse response of a linear system with the input $Y(\cdot)$ and the output $\hat{X}(\cdot)$ minimizing

$$E[X(t_1) - \hat{X}(t_1|t)]' W[X(t_1) - \hat{X}(t_1|t)] = \| X(t_1) - \hat{X}(t_1|t)\|^2_{\text{q.m.}} \quad (4.98)$$

where W is any $n \times n$ positive semi-definite matrix (it can be shown that the minimization of (4.98) is independent of W.

The state estimation problem can be divided into three classes: (1) filtering if $t = t_1$, (2) prediction if $t_1 > t$, (3) smoothing if $t_1 < t$.

Filtering

The optimal solution is given in Kalman's original work. We know $\hat{X}(t|t)$ is the optimal solution if and only if it satisfies:

$$E[X(t) - \hat{X}(t|t)] Y'(\tau) = 0, \quad \text{for } 0 \leq \tau \leq t \quad (4.99)$$

which is the orthogonality principle; without any loss of generality we have assumed $t_0 = 0$.

Since we expect the optimal solution to be a combination of the $\hat{X}(\cdot)$ and the measurement $Y(t)$, we make a guess that $\hat{X}(t|t)$ is the solution of the differential equation

$$\dot{\hat{X}} = F_1(t) \hat{X}(t) + F_2(t) Y(t), \quad \hat{X}(0|0) = 0 \quad (4.100)$$

where $F_1(t)$ and $F_2(t)$ are chosen such that the orthogonality condition in (4.99) is satisfied. We know that if the orthogonality condition is satisfied the solution must be optimal (unique). Thus, if $F_1(t)$ and $F_2(t)$ could be found such that Eq. (4.99) is satisfied, then the $\hat{X}(\cdot)$ corresponding to these $F_1(t)$ and $F_2(t)$ must be optimal.

Indeed, it can be shown that the solution of the form given by Eq. (4.100) satisfies the orthogonality principle. The solution is quite tedious. Let us state the results via the theorem (the proof is given later).

Theorem 7

The optimal K-B filtering estimate $\hat{X}(t)$ is the solution of Eq. (4.100), where

$$F_1(t) = [A(t) - F_2(t) C(t)] \qquad (4.101)$$

and

$$F_2(t) = P(t) C'(t) L^{-1}(t) \qquad (4.102)$$

$L(t)$ is given by Eq. (4.97) and $P(t)$ is given by:

$$P(t) = E\left\{ [X(t) - \hat{X}(t|t)] \, [X(t) - \hat{X}(t|t)]' \right\} \qquad (4.103)$$

and can be obtained as the solution of the nonlinear differential equation

$$\dot{P} = AP + PA' - PC' L^{-1} CP + BQB' \qquad (4.104)$$

with the given initial condition $P(0) = E(X(0|0) \, X'(0|0))$. Note that we have dropped the argument t for convenience. The proof will be given later, but we shall first give an example, after some discussion.

In Figure 4-4, the optimum continuous filter is diagrammed. The input to the system is the observation $Y(t)$ which is the contaminated signal and the outputs could be considered as $\hat{X}(t|t)$ or $C\hat{X}(t|t)$, where $C\hat{X}(t|t)$ is the optimal estimate to $Y(t)$.

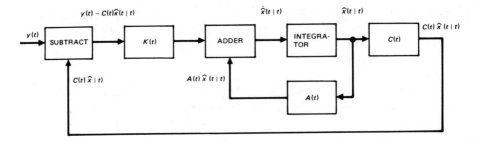

Fig. 4-4. Optimum Continuous Filter

Example 9

An object moves with an unknown constant velocity V on a straight line trajectory. Suppose we observe the projectile at the initial time $t_0 = 0$ at a known point $s(0)$ as shown by:

Thereafter the projectile is tracked for τ seconds. The observation consists of the displacement from the origin which has been contaminated by additive white noise of spectral density N_0 watts/hertz. Let us find the Kalman filter yielding the optimal linear estimate of V. Assume the velocity V has the variance σ^2.

Solution

Since the speed is constant $\dot{V} = 0$ and the observation $\overline{Y}(t)$ by definition is:

$$\overline{Y}(t) = s(t) + n(t) = s(0) + tV + n(t)$$

If we let $Y(t) = \overline{Y}(t) - s(0)$, the dynamic system becomes

$$\dot{V} = 0$$

$$Y(t) = tV + n(t)$$

Thus, $A = B = 0$, $C = t$, and $L = N_0$ from which

$$\dot{\hat{V}} = F_2(t)[Y(t) - t\hat{V}]$$

$$\dot{P} = -P^2(t)\, t^2/N_0$$

$$F_2(t) = P(t)\, t/N_0$$

The initial conditions are $V(0 \mid 0) = 0$, $P(0) = \sigma^2$.

To solve for $\hat{V}(t)$,* we need to obtain $F_2(t)$ which in turn requires the solution of $P(t)$:

$$\int_{P(0)=\sigma^2}^{P(t)} P^{-2}\, dP = -\frac{1}{N_0} \int_0^t \tau^2\, d\tau$$

from which

$$P(t) = \frac{3N_0 \sigma^2}{3N_0 + \sigma^2 t^3}$$

Thus,

$$\hat{V}(t) = \int_0^t \frac{3\sigma^2 \zeta}{3N_0 + \sigma^2 \zeta^3} [Y(\zeta) - \zeta \hat{V}(\zeta)]\, d\zeta, \quad 0 \leqslant t \leqslant T \quad (4.105)$$

For the special case that $\hat{V}(T) \to$ constant as $T \to \infty$, from (4.96) we get:

$$\frac{3\sigma^2 \zeta}{3N_0 + \sigma^2 \zeta^3} [y(\zeta) - \zeta \hat{V}] \approx 0, \quad \text{for large } \zeta$$

---

*We shall denote $\hat{V}(t,t)$ by $\hat{V}(t)$.

Remark 5. In filtering we shall often write $\hat{V}(t)$ instead of $\hat{V}(t|t)$ or

$$\hat{V}(T) \approx \frac{VT + n(T) - s(0)}{T} = V + \frac{n(T) - s(0)}{T}$$

which implies

$$\hat{V}(T) \to V, \quad \text{as } T \to \infty$$

That is, if the contaminated signal is observed for a long time, we should get the exact estimate.

Example 10

Let the observation $Y(t)$ be given by:

$$Y(t) = d \cos(\omega_0 t - \theta_0) + v(t) \tag{4.106}$$

where d, ω_0, θ_0 are, respectively, the amplitude, carrier frequency, and phase. Let $v(t)$ be a white noise process with a variance of unity. Assume that ω_0, θ_0 are known exactly. Estimate d.

Solution

Since d is constant, then $\dot{d} = 0$. Now, we can have:

$$\dot{X} = 0$$

$$X(0) = X(t) = d$$

$$Y(t) = \cos(\omega_0 t - \theta_0) X + v(t)$$

Hence, $A = B = 0$, $C = \cos(\omega_0 t - \theta_0)$, $Q = 1$ and $L = 1$. From Eqs. (4.100)–(4.103):

$$\dot{P} = -[\cos(\omega_0 t - \theta_0) P(t)]^2 \tag{4.107}$$

$$F_1(t) = -F_2(t) C = -[\cos(\omega_0 t - \theta_0)]^2 P(t) \tag{4.108}$$

Thus,

$$\dot{\hat{X}} = F_1(t)\hat{X}(t) + F_2(t)Y(t) \qquad (4.109)$$

where $F_1(t)$ and $F_2(t)$ are given by the previous equations.

The solution $\hat{X}(t)$ requires the solution $P(t)$ from Eq. (4.107). It is apparent that even for the scalar case, the solution can become fairly tedious.

Remark 6. Note that $\hat{X}(t)$ is the estimate of $X(t)$, given the observation $Y(t)$. The corresponding uncertainty (covariance) of $\hat{X}(t)$ is given by $P(t)$. Since,

$$P(t) = E[ee'] = E\left\{[X(t) - \hat{X}(t)][X(t) - \hat{X}(t)]'\right\}$$

$$= E\left\{[\hat{X}(t) - X(t)]'[\hat{X}(t) - X(t)]\right\}$$

$$= E\left\{[\hat{X}(t) - E\hat{X}(t)][\hat{X}(t) - E\hat{X}(t)]'\right\} = \operatorname{cov}\hat{X}(t)$$

for the case of the unbiased estimate, then $P(t)$ is indeed a covariance.

Example 11

In the previous example suppose d is known perfectly and it is desired to estimate ω_0, and θ_0. Obtain the model and the form of the solution.

Solution

Let

$$X(t) = d\cos(\omega_0 t - \theta_0)$$

Then

$$\dot{X}(t) = -d\omega_0 \sin(\omega_0 t - \theta_0)$$

Now if we define $X_1(t) = X(t)$ and $X_2(t) = \dot{X}_1(t) = \dot{X}(t)$, we have:

$$\dot{X}_1 = -d\omega_0 \sin(\omega_0 t - \theta_0) = X_2(t)$$

$$\dot{X}_2 = -d\omega_0^2 \cos(\omega_0 t - \theta_0) = -\omega_0^2 X_1(t)$$

so that

$$\frac{d}{dt}\begin{bmatrix} X_1 \\ X_2 \end{bmatrix} = \begin{bmatrix} 0 & 1 \\ -\omega_0^2 & 0 \end{bmatrix} \begin{bmatrix} X_1 \\ X_2 \end{bmatrix} \quad (4.110)$$

$$Y(t) = \begin{bmatrix} 1 & 0 \end{bmatrix} \begin{bmatrix} X_1 \\ X_2 \end{bmatrix} + v(t) \quad (4.111)$$

Thus, by inspection:

$$A = \begin{bmatrix} 0 & 1 \\ -\omega_0^2 & 0 \end{bmatrix}$$

$$B = 0$$

$$C = \begin{bmatrix} 1 & 0 \end{bmatrix}$$

$$D = 1$$

Now the solution is more involved and the estimate $\hat{X}(t)$ of $X(t)$ with its covariance $P(t)$ can be obtained as before.

Example 12

This example is taken from reference [8]. Assume that $Y(t)$ is a white noise process with unknown mean X. Thus,

$$EY(t) = X \quad (4.112a)$$

$$E\left\{[Y(t_1) - X][Y(t_2) - X]'\right\} = L\delta(t_2 - t_1) \quad (4.112b)$$

Suppose we want to estimate X when the observation $Y(t)$ is received over the interval $[0, t]$.

Solution

Since X is constant, we can construct a model as follows.

$$\dot{X} = 0$$

$$Y = X(t) + v(t)$$

$$Ev(t_2)v(t_1) = L\,\delta(t_2 - t_1)$$

From Eq. (4.104), we get:

$$\dot{P}(t) = -L^{-1} P^2(t)$$

or

$$\frac{\dot{P}(t)}{P^2(t)} = -L^{-1}$$

Integrating both sides yields:

$$-P^{-1}(t) = -L^{-1} t + c \longleftrightarrow P(t) = \frac{1}{L^{-1} t - c}$$

where c is a constant.

However, at $t = 0$, we get:

$$P(0) = \frac{1}{c} \longleftrightarrow c = \frac{1}{P(0)}$$

Thus,

$$P(t) = \frac{1}{L^{-1} t + 1/P(0)} \qquad (4.113)$$

Now, substituting Eq. (4.113) into Eq. (4.100), i.e.,

$$\dot{\hat{X}} = F_1(t)\hat{X}(t) + F_2(t)Y(t), \quad \hat{X}(0) = 0$$

where, from Eqs. (4.102) and (4.103),

$$F_1(t) = A(t) - F_2(t)C(t) = \frac{-1}{t + \frac{L}{P(0)}}$$

$$F_2(t) = P(t)C'L^{-1}(t) = \frac{1}{t + \frac{L}{P(0)}}$$

we obtain:

$$\dot{\hat{X}} = \frac{-1}{t + \frac{L}{P(0)}}\hat{X}(t) + \frac{1}{t + \frac{L}{P(0)}}Y(t), \quad \hat{X}(0) = 0 \quad (4.114)$$

From the above equation, the transition matrix $\Phi(t,0)$ is given by:

$$\Phi(t,0) = \frac{\frac{L}{P(0)}}{t + \frac{L}{P(0)}} \quad (4.115)$$

This is true because

$$\dot{\Phi} = \frac{-1}{t + \frac{L}{P(0)}}\Phi, \quad \Phi(0,0) = 1$$

Equation (4.14) can be solved by using Eq. (4.90). Thus,

$$\hat{X}(t) = \int_0^t \overbrace{\frac{\frac{L}{P(0)}}{t + \frac{L}{P(0)}} \frac{\tau + \frac{L}{P(0)}}{\frac{L}{P(0)}}}^{\Phi(t,\tau)} \frac{1}{\tau + \frac{L}{P(0)}} Y(\tau) d\tau$$

Simplification of the above gives rise to:

$$\hat{X}(t) = \frac{1}{t + \frac{L}{P(0)}} \int_0^t Y(\tau)\,d\tau \qquad (4.116)$$

Since both L and $P(0)$ are constants, we obtain:

$$\hat{X}(t) = \lim_{t \to \infty} \frac{1}{t} \int_0^t Y(\tau)\,d\tau$$

which is expected. Thus, for a long observation, $\hat{X}(t)$ becomes independent of $P(0)$.

Before proving Theorem 7 it would be helpful to discuss some of its important properties.

Discussion

It is obvious that K-B filtering is a boundary value problem. The linear estimate $\hat{X}(t|t)$ is given as the solution of Eq. (4.100). The solution is weighted on both the estimate and the measurement with the appropriate weighting factors $F_1(t)$ and $F_2(t)$ given by Eqs. (4.101) and (4.102), respectively. The quantities $F_1(t)$ and $F_2(t)$ depend on the solution of $P(t)$, which is obtained from the nonlinear differential equation given by Eq. (4.104). Note that the solution of $P(t)$ obtained by Eq. (4.104) depends on the known parameters A, C, and the second order statistics of U and v. Consequently, $P(t)$, $F_1(t)$, and $F_2(t)$ can be determined before any measurement is received. In general, the best m.s.e. requires the knowledge of the first and the second moments of the random vectors X and Y and that their entire probability density function. Therefore, any two random vectors which have identical means and covariances yield the same estimator $\hat{X}$.

Definition 1 introduced the concept of a conditional estimator. Another concept that will be used in the sequel is the unconditional unbiased estimator, which is defined below.

Definition 5

An estimate $\hat{X}$ is said to be an unconditional unbiased estimate if

$$E_Y \hat{X} = EX.$$

It can be shown that the m.s.e. $\hat{X}$ is an unconditional unbiased estimate since:

$$E[X - \hat{X}] = 0$$

It can also be verified that

$$\widehat{A_1 X + A_2} = A_1 \hat{X} + A_2$$

for nonstochastic matrices A_1 and A_2.

Now we shall prove Theorem 7; see reference [12].

Proof

We can extend the general Wiener-Hopf equation given by (4.48) to the case where the signal $s(t)$ is changed to the vector X. Then the cross correlation function $R_{sY}(t - \alpha)$ will be simply changed to $R_{XY}(t - \alpha)$. Let us also assume that the mean of X and Y is not zero. Then we will change $R_{XY}(t - \alpha)$, and $R_Y(\sigma - \alpha)$ to $C_{XY}(t - \alpha)$ and $C_Y(\sigma - \alpha)$, respectively. Thus, the generalized Wiener-Hopf equation becomes:

$$C_{XY}(t - \alpha) = \int_{t_0}^{t} G(t, \sigma) C_Y(\sigma - \alpha) d\sigma \qquad (4.117)$$

where $G(t, \sigma)$ is the generalized impulse response.

The above equation is equivalent to the orthogonality condition. Let us take the left-hand side partial derivate of $C_{XY}(\cdot)$ to get:

$$\frac{\partial}{\partial t} C_{XY}(t - \alpha) = \frac{\partial}{\partial t} E\left\{[X(t) - E[X(t)]] [Y(\alpha) - E[Y(\alpha)]]'\right\}$$

$$= E\left[\frac{\partial}{\partial t} X(t) Y'(\alpha)\right] - E\left[\frac{\partial X}{\partial t}\right] E[Y'(\sigma)]$$

$$= E\left[(AX + BU) Y'(\sigma)\right] - E[AX + BU] E[Y'(\sigma)]$$

$$= A(t) C_{XY}(t - \alpha) + B(t) C_{UY}(t - \alpha) \qquad (4.118)$$

The above equation was obtained by using Eq. (4.95). Since $U(t)$ is independent of both $v(\alpha)$ and $X(\alpha)$ for $\alpha < t$. Thus, $C_{UY}(t - \alpha) = 0$. On the other hand, the derivative of the right-hand side of Eq. (4.117) yields:

$$\frac{\partial}{\partial t} \int_{t_0}^{t} G(t,\sigma) C_Y(\sigma - \alpha) d\alpha = \int \frac{\partial G(t,\sigma)}{\partial t} C_Y(\sigma - \alpha) d\sigma$$

$$+ G(t,t) C_Y(t - \alpha) \quad (4.119)$$

However, the left-hand side of (4.119) after denoting $Z(t)$ for $C(t)X(t)$, can be written as:

$$\frac{\partial}{\partial t} \int_{t_0}^{t} G(t,\sigma) C_Y(\sigma - \alpha) d\sigma$$

$$= \frac{\partial}{\partial t} \int_{t_0}^{t} G(t,\sigma) E \left\{ [Y(\sigma) - m_Y] [Y(\alpha) - m_Y]' \right\} d\sigma$$

$$= \frac{\partial}{\partial t} \int_{t_0}^{t} G(t,\sigma) E \left\{ [Z(\sigma) - m_Y + v(\sigma)] [Z(\alpha) - m_Y + v(\alpha)]' \right\} d\sigma$$

$$= \frac{\partial}{\partial t} \int_{t_0}^{t} G(t,\sigma) C_Z(\sigma - \alpha) d\sigma + \frac{\partial}{\partial t} G(t,\alpha) L(\alpha)$$

$$= \int_{t_0}^{t} \frac{\partial G(t,\sigma)}{\partial t} C_Z(\sigma - \alpha) d\sigma + G(t,t) C_Z(t - \alpha) + \frac{\partial G(t,\alpha)}{\partial t} L(\alpha)$$

$$(4.120)$$

Now, using $Y(t) = C(t) X(t) + v(t) = Z(t) + v(t)$ and the fact that $C_{UY}(t - \alpha) = 0$, following Eq. (4.119), we can obtain:

$$C_Z(t - \alpha) = E[Z(t) Z'(\alpha)] = A(t) C_{XY}(t - \alpha)$$

$$= A(t) \int_{t_0}^{t} G(t,\sigma) C_Y(\sigma - \alpha) d\sigma \qquad (4.121)$$

Now, if we combine (4.117), (4.118), (4.119), (4.120), and (4.121), we obtain:

$$\int_{t_0}^{t} \left[A(t)G(t,\sigma) - \frac{\partial A(t,\sigma)}{\partial t} - G(t,t)A(t)G(t,\sigma) \right] C_Y(\sigma - \alpha) d\sigma = 0, \; t_0 \leqslant \alpha < t$$

$$(4.122)$$

Then, from the above:

$$A(t)G(t,\sigma) - \frac{\partial A(t,\sigma)}{\partial t} - G(t,t)A(t)G(t,\sigma) = 0, \; t_0 \leqslant \sigma \leqslant t \qquad (4.123)$$

Since

$$\hat{X}(t) = \int_{t_0}^{t} G(t, \sigma) Y(\sigma) d\sigma \qquad (4.124)$$

for the optimal solution, combining this with (4.123) yields

$$\dot{\hat{X}} = \int_{t_0}^{t} \frac{\partial}{\partial t} G(t,\sigma) Y(\sigma) d\sigma + G(t,t) Y(t)$$

$$= \int_{t_0}^{t} [A(t) G(t,\sigma) - G(t,t) C(t) G(t,\sigma) Y(\sigma)] d\sigma + G(t,t) Y(t)$$

which implies

$$\dot{\hat{X}} = A(t)\,\hat{X}(t) + G(t,t)\,[Y(t) - C(t)]\,\hat{X}(t)$$

$$= [A(t) - G(t,t)\,C(t)]\,\hat{X}(t) + A(t,t)\,Y(t)$$

Thus,

$$F_1(t) = [A(t) - G(t,t)\,C(t)] = [A(t) - F_2(t)\,C(t)]$$

This part of the proof is done. Note that $F_2(t) = G(t,t)$.

It can be shown (left as an exercise) that:

$$\frac{de}{dt} = [A(t) - F_2(t)\,C(t)]\,e(t) + B(t)\,U(t) - F_2(t)\,v(t) \qquad (4.125)$$

and

$$C_{XY}(t - \alpha) = C_{XZ}(t - \alpha) \qquad (4.126)$$

where

$$e(t|t) = X(t) - \hat{X}(t)$$

We can also obtain $C_Y(\sigma - \alpha)$ as:

$$C_Y(\sigma - \alpha) = E\left\{Y(\sigma)\,Y'(\alpha)\right\}$$

$$= E\left\{[Z(\sigma) + v(\sigma)]\,[Z(\alpha) + v(\alpha)]'\right\}$$

$$= C_Z(\sigma - \alpha) + C_v(\sigma - \alpha)$$

$$= C_Z(\sigma - \alpha) + L(\sigma)\,\delta(\sigma - \alpha) \qquad (4.127)$$

Thus,

$$C_{XY}(t - \alpha) = \int_{t_0}^{t} G(t, \sigma) C_Y(\sigma - \alpha) d\sigma =$$

$$C_{XZ}(t - \alpha) = \int_{t_0}^{t} [G(t, \sigma) C_Z(\sigma - \alpha) d\sigma + G(t, \alpha) L(\alpha)$$

from which

$$G(t, t) L(t) = E\{[X(t) - \hat{X}(t)] Y'(t)\} = C_{eX}(0|t) C'(t)$$

If we let

$$P(t) \triangleq C_{eX}(0|t)$$

and since $L^{-1}(t)$ exists (assumed to be positive semi-definite), we can obtain:

$$F_2(t) = G(t, t) = P(t) C'(t) L^{-1}(t) \qquad (4.128)$$

The only thing needed in the proof is to solve for $P(t)$. From Eq. (4.125), let us solve for $e(t)$ or, equivalently, $e(t|t)$.

Let $\hat{\Phi}(t, \tau)$ denote the transition matrix of (4.125), then

$$e(t) = \hat{\Phi}(t, t_0) + \int_{t_0}^{t} \hat{\Phi}(t, s) [-F_2(s) + v(s) + BU(s)] ds$$

Substituting

$$P(t) = E[e(t) e'(t)]$$

in the above, then, with the assumption that $e(0)$ and $U(s)$ and $v(s)$ are uncorrelated, we obtain (after some manipulation):

$$P(t) = \hat{\Phi}(t,t_0) P(t_0) \hat{\Phi}'(t,t_0) + \int_{t_0}^{t} \hat{\Phi}(t,s) [F_2(s) L(s) F_2'(s)$$

$$+ BQB'] \hat{\Phi}'(t,s) ds$$

Thus, upon differentiation, we obtain:

$$\dot{P} = [A - F_2(t) C'] P(t) + P(t) [A' - C' F_2'(t)]$$

$$+ F_2(t) L F_2'(t) + BQB'$$

where we have used

$$\frac{d}{dt} \hat{\Phi} = [A - F_2(t) C(t)] \hat{\Phi}(t,t_0)$$

in the above.

Now, if we substitute $F_2(t)$ from (4.128), we shall obtain the result, i.e.,

$$\dot{P} = AP + PA' - PC' L^{-1} CP + BQB'$$

which completes the proof.

Remark 7. Very often $F_2(t)$ is changed notationally to $K(t)$ and the gain $F_1(t)$ to $F(t)$; see (4.101). Then, Eq. (4.100) can be rewritten as:

$$\dot{\hat{X}} = [A(t) - K(t) C(t)] \hat{X}(t) + K(t) Y(t)$$

$$= A(t) \hat{X}(t) + K(t) [Y(t) - C(t) \hat{X}(t)] \qquad (4.129)$$

4.10.2 Prediction

The solution of the prediction problem is a simple extension of the filtering problem, and it is actually presented by Kalman in his initial paper.

Suppose we wish to estimate $X(t_1)$ based on the observation $Y(t)$ given on the interval $0 \leq \tau \leq t$ for $t_1 > t$. The solution $\hat{X}(t_1|t)$ is given by:

$$\hat{X}(t_1|t) = \Phi(t_1, t)\hat{X}(t|t), \quad t_1 > t \qquad (4.130)$$

where $\Phi(\cdot,\cdot)$ is the transition matrix corresponding to Eq. (4.100).

The covariance matrix is found accordingly. Therefore, for prediction problems, we must first obtain a filtered estimate of the state, up to the range of available data.

Thus, $Y(\lambda)$ should be set equal to zero for $\lambda > t$, and $\hat{X}(t)$ serves as the initial condition in Eq. (4.90).

4.10.3 Smoothing

In smoothing $0 \leq t_1 < t_2$, where it is desired to estimate $X(t_1)$, given an observation over the interval $0 \leq \tau < t_2$. The smoothing problem, at first glance seems to be identical to the prediction problem, but smoothing is far more involved than either filtering or prediction. This is because $U(\cdot)$ and $Y(\cdot)$ are not uncorrelated over the range of interest. Thus the result given by (4.130) cannot hold. In reference [21], the smoothing problem is divided into three separate areas:

(1) Fixed-interval smoothing, where the initial and final times are fixed and we seek an estimator $\hat{X}(t|t_2)$, where t varies over the interval $[0, t_2]$.

(2) Fixed-point smoothing, where for the estimator $\hat{X}(t|t_2)$, t is fixed and $0 \leq t \leq t_2$.

(3) Fixed-lag smoothing, where both t_1 and t_2 vary, but $t_2 = t_1 + T$, such that T is fixed and $\hat{X}(t_2 - T|t_2)$ is sought.

There are three different solutions for these three classes of smoothing. One of the most interesting approaches to solving the smoothing problem is due to Kailath and Frost, references [24]-[26], who use the "innovation" approach. Another interesting solution for fixed-interval smoothing is due to Fraser [23] and Fraser and Potter [33]. They showed that the optimal smoother can be considered to be the combination of two uncorrelated optimal filter estimates, one running forward in time and the other backward. In this book we shall concern ourselves with the solution of the continuous smoothing problem. A special case of smoothing will be discussed in the discrete version.

4.10.4 Discrete Kalman Recursive Estimation

In Subsections 4.10.1-4.10.3, we have discussed the continuous model representing continuous random processes. We shall now begin the discussion of the discrete-time version of the problem, because this version must be used for computer implementation. There are a number of inherent advantages; for example, the discrete algorithms can be manipulated by hand and the step-by-step processing of information lends itself to a simple development.

In subsequent sections we shall discuss prediction, filtering, and smoothing.

4.10.5 One-Step Prediction

Consider the discrete dynamic system:

$$X(k+1) = \bar{A}X(k) + \bar{B}U(k) \qquad (4.131)$$

$$Y(k) = \bar{C}X(k) + v(k) \qquad (4.132)$$

The signal and noise have the following statistical moments:

$$EU(k) = Ev(k) = 0 \qquad (4.133a)$$

$$EU(k_1)\,U'(k_2) = \bar{Q}\Delta(k_2 - k_1) \qquad (4.133b)$$

$$Ev(k_1)\,v'(k_2) = \bar{L}\Delta(k_2 - k_1) \qquad (4.133c)$$

$$EU(k_1)\,v'(k_2) = 0 \qquad (4.133d)$$

where $\bar{A}, \bar{B}, \bar{C}, \bar{Q}$, and $\bar{L}$ are $n \times n$, $n \times p$, $q \times n$, $p \times p$, and $q \times q$ matrices, respectively, which are in general a function of k. The quantity $\Delta(k_2 - k_1)$ is defined as follows:

$$\Delta(k_2 - k_1) = \begin{cases} 1, & \text{if } k_1 = k_2 \\ 0, & \text{otherwise} \end{cases} \qquad (4.134)$$

Q and L are assumed to be positive definite.

The initial state $X(0)$ is assumed to be a random vector with a known *a priori* covariance matrix $\bar{P}(0)$.

We would like to find the estimate of the vector $X(k + 1)$ denoted as $\widehat{X}(k + 1)$, which is a linear function of $Y(0), Y(1), \ldots, Y(k)$ minimizing:

$$E[X(k + 1) - \widehat{X}(k + 1]' W[X(k + 1) - \widehat{X}(k + 1)] \quad (4.135)$$

where W is any positive semi-definite matrix; for example $W = I$ is a proper choice, and it can be shown that the optimal solution is independent of the choice W.

The solution to this problem can be obtained by conjecturing that the estimator has the form:

$$\widehat{X}(k + 1) = F_1(k) \widehat{X}(k) + F(k) Y(k) \quad (4.136)$$

If $\widehat{X}(k + 1)$ satisfies the orthogonality principle for some F_1 and F, then it would be optimal (see Theorem 2). Consequently, we are making a guess that our optimal solution would be of the form (4.136), where F_1 and F are to be determined in such a way that the orthogonality principle would be satisfied. From Eq. (4.136) it can be seen that the one-step predictor $\widehat{X}(k + 1)$ is weighted on the previous estimate $\widehat{X}(k)$ and its corresponding measurement $Y(k)$ with the weighting factors F_1 and F, respectively. The solution of the optimal estimator is given via Theorem 8; see references [8], [21], and [24]-[26].

Theorem 8

The one-step predictor $\widehat{X}(k + 1)$ given by Eq. (4.136) is optimal iff the matrices F_1 and F will satisfy the relation:

$$F_1(k) = \overline{A} - F(k) \overline{C} \quad (4.137)$$

$$F(k) = \overline{AP(k)} \overline{C}' [\overline{CP(k)} \overline{C}' + \overline{L}]^{-1} \quad (4.138)$$

where $\overline{P}(k)$ is defined by

$$E[X(k) - \widehat{X}(k)] [X(k) - \widehat{X}(k)]' \quad (4.139)$$

and satisfies the following equation:

$$\overline{P}(k + 1) = [\overline{A} - F(k) \overline{C}] \overline{P}(k) [\overline{A} - F(k) \overline{C}]' + \overline{BQB'} + F(k) \overline{L} F(k)' \quad (4.140)$$

The initial conditions of Eqs. (4.136) and (4.140) are given by $\hat{X}(0)$ and $\bar{P}(0)$, respectively.

Proof

As mentioned above, one can prove the theorem via the orthogonality principle; see reference [8] for the proof. However, we shall prove the theorem by a different approach [24]. Without any loss of generality, we may assume that $\hat{X}(0) = EX(0) = 0$. Let $\tilde{Y}(\cdot)$ be defined by:

$$\tilde{Y}(k) \triangleq Y(k) - \bar{C}(k)\hat{X}(k) \tag{4.141}$$

where $\tilde{Y}(k)$ is called the *innovation* process.

Since $\hat{X}(k)$ represents a linear estimator, one can write (left as an exercise):

$$\hat{X}(k) = \sum_{i=0}^{k-1} F_k^i \tilde{Y}(k) \tag{4.142}$$

where F_k^i's are the weighting factors (gains) to be determined.

It can be shown that the innovation process $\tilde{Y}(k)$ is uncorrelated with the sequence $Y(0), Y(1), \ldots, Y(k-1)$ and thus

$$E\{[X(k) - \hat{X}(k)]\,\tilde{Y}'(j)\} = 0, \quad 0 \leq j \leq k-1 \tag{4.143}$$

Substituting Eq. (4.142) into Eq. (4.143) will yield:

$$E[X(k)\tilde{Y}'(j)] = \sum_{i=1}^{k-1} F_k^i E[\tilde{Y}(i)\tilde{Y}'(i)], \quad 0 \leq j \leq k-1 \tag{4.144}$$

Noting the fact that the innovation process $\tilde{Y}(j)$ is a white noise term with

$$E[\tilde{Y}(j)\tilde{Y}'(i)] = [\bar{C}(i)\bar{P}(i)\bar{C}'(i) + \bar{L}(i)]\,\Delta_{ij} \tag{4.145}$$

and verifying that:

$$E[X(k)\tilde{Y}'(j)] = \bar{A}(k-1)\bar{A}(k-2)\ldots\bar{A}(j)\bar{P}(j)\bar{C}'(j) \tag{4.146}$$

it can be shown that (the details are left as an exercise):

$$\hat{X}(k+1) = \sum_{j=0}^{k} F_{k+1}^j \tilde{Y}(j) = [\bar{A}(k) - F(k)\bar{C}(k)]\hat{X}(k) + F(k)Y(k)$$

where $F(k)$ is given by

$$F(k) = \bar{A}(k)\bar{P}(k)\bar{C}'(k)[\bar{C}(k)\bar{P}(k)\bar{C}'(k) + \bar{L}(k)]^{-1}.$$

The last two equations coincide with Eqs. (4.136) and (4.138), respectively. The verification of Eq. (4.140) is conceptually very simple, but requires substantial algebraic manipulation. To verify Eq. (4.140), we need to calculate the covariance of $X(k+1) - \hat{X}(k+1)$, which is $\bar{P}(k+1)$, and substitute Eqs. (4.131)-(4.133) into the equation. This would complete the proof.

The problem of predicting more than one step is a simple extension of the above. For example, $\hat{X}(k+j)$ for $j > 1$, can be obtained as

$$\hat{X}(k+j) = \bar{A}^{j-1}\hat{X}(k+1) \tag{4.147}$$

and the associated covariance matrix is found accordingly.

4.10.6 Discrete Filtering

The filtering problem is the determination of the estimate of $X(k)$ given the observations $Y(0), Y(1), \ldots, Y(k)$. Let us denote the filtered value of $X(k)$ by $\hat{X}^0(k)$. It can be shown that $\hat{X}^0(k)$ is given by:

$$\hat{X}^0(k) = [\bar{A}(k)]^{-1}\hat{X}(k+1) \tag{4.148}$$

where $\hat{X}(k+1)$ is determined from Eqs. (4.136)-(4.140). By utilizing these equations we obtain:

$$\hat{X}^0(k) = [I - F(k)\bar{C}]\bar{A}\hat{X}^0(k-1) + F(k)Y(k) \tag{4.149a}$$

$$F(k) = \bar{P}(k)\bar{C}'[\overline{CP}(k)\bar{C}' + \bar{L}]^{-1} \tag{4.149b}$$

$$\bar{P}(k+1) = \bar{A}(I - F(k)\bar{C}]\bar{P}^0(k)\bar{A}' + \overline{BQB}' \tag{4.149c}$$

$$\bar{P}^0(k+1) = \bar{P}(k+1) - F(k+1)\bar{C}(k+1)\bar{P}(k+1) \tag{4.149d}$$

which is the solution yielding the optimal filter. The matrices $\bar{A}, \bar{B}, \bar{C},$ and $\bar{Q}$ are all functions of k and $\bar{P}^0(k)$ is the covariance of $X(k) - \hat{X}^0(k)$. The derivation of Eqs. (4.148) and (4.149) will be discussed following the section on smoothing (interpolation). Because the smoothing problem can sometimes be considered as a combination of two estimators, we shall first discuss the combination process.

4.11 COMBINATION OF UNBIASED ESTIMATORS

Suppose we are given two unbiased estimates $\hat{X}_1(t)$ and $\hat{X}_2(t)$ of the same state $X(t)$. There are two cases to consider: either $\hat{X}_1$ and $\hat{X}_2$ are correlated or they are uncorrelated. We shall discuss both cases below.

4.11.1 The Estimates are Uncorrelated

$\hat{X}_1$ and $\hat{X}_2$ are said to be uncorrelated if

$$E[X - \hat{X}_1][X - \hat{X}_2]' = 0 \tag{4.150}$$

If $E[X - \hat{X}_1][X - \hat{X}_2]' \neq 0$, then $\hat{X}_1$ and $\hat{X}_2$ are uncorrelated. The optimal estimate of X is obtained as follows:

$$\hat{X} = P(P_1^{-1}\hat{X}_1 + P_2^{-1}\hat{X}_2) \tag{4.151}$$

$$P = (P_1^{-1} + P_2^{-1})^{-1} \tag{4.152}$$

where P_i is defined for $i = 1, 2$ by:

$$P_i = E(X - \hat{X}_i)(X - \hat{X}_i)'$$

To prove Eqs. (4.151) and (4.152), we shall seek a solution $\hat{X}$ of the form:

$$\hat{X} = L_1 \hat{X}_1 + L_2 \hat{X}_2 \tag{4.153}$$

where L_1 and L_2 are to be determined. Since $\hat{X}_1$, $\hat{X}_2$, and $\hat{X}$ are unbiased (conditional or unconditional), then it is obvious that:

$$E\hat{X} = (L_1 + L_2)EX$$

which implies that:

$$L_1 + L_2 = I \tag{4.154}$$

Calculating the covariance of $X - \hat{X}$ and using Eqs. (4.153)–(4.154) gives rise to:

$$P = L_1 P_1 L_1' + (I - L_1)P_2(I - L_1)' \tag{4.155}$$

Minimizing the m.s.e. $E\|X - \hat{X}\|^2 = E[(X - \hat{X})'(X - \hat{X})]$ with respect to L_1 is equivalent to taking the derivative of P with respect to L_1 and setting it equal to zero (why?). This yields:

$$L_1 = P_2(P_1 + P_2)^{-1} \tag{4.156}$$

$$L_2 = P_1(P_1 + P_2)^{-1} \tag{4.157}$$

which upon substitution into Eq. (4.155) gives us Eq. (4.152). Using Eq. (4.152) in Eq. (4.153) yields Eq. (4.151), which concludes our assertions.

4.11.2 The Estimates are Correlated

The solution for correlated estimators is given by:

$$\hat{X} = L_1 \hat{X}_1 + L_2 \hat{X}_2$$

$$P = L_1 P_1 L_1' + L_2 P_2 L_2' + L_1 P_{12} L_2' + L_2 P_{12} L_1' \tag{4.158}$$

where

$$P_{12} = E[X - \hat{X}_1][X - \hat{X}_2]'$$

L_1 and L_2 can be obtained in a way similar to the previous case.

4.12 DISCRETE SMOOTHING

Discrete smoothing, as with its continuous counterpart, can be divided into three categories. We shall deal with the special case, where we seek the best estimate of the state $\hat{X}(k - j)$ for some $j > 0$, given the observation $Y(0)$, $Y(1)$, $Y(2)$, ..., $Y(k)$. Let us denote the estimate by $\hat{X}^{(j)}(k - j)$. The solution of the smoothing problem is given below as

$$\hat{X}^{(j)}(k + j) = R(k - j)\hat{X}^{(j-1)}(k - j + 1)$$

$$+ [(\overline{A})^{-1}(k - j) - R(k - j)]\hat{X}(k - j + 1) \tag{4.159}$$

where

$$R(i) = (\overline{A(i)})^{-1}[I - \overline{B}(k)\overline{QB}'(i)(\overline{P}(i + 1)^{-1}] \tag{4.160}$$

and $\overline{P}(i)$ satisfies Eq. (4.140).

The computation given by Eqs. (4.159) and (4.160) are carried out as follows.

Having received the observations $Y(0), Y(1), \ldots, Y(k)$, the results of one-step prediction yield the value $\hat{X}(k+1)$. The values $\hat{X}(k), \hat{X}(k-1), \ldots, \hat{X}(k-j+1)$ are stored along with the values $\overline{P}(k), \overline{P}(k-1), \ldots, \overline{P}(k-j+1)$. Using these values and starting with $\hat{X}^0(k)$ given by Eq. (4.149) and utilizing Eqs. (4.159)–(4.160) we can recursively obtain $\hat{X}^{(1)}(k-1), \hat{X}^{(2)}(k-2), \ldots, \hat{X}^{(j)}(k-j)$, which is the desired result. For example, to obtain the best estimate of $X(10)$ based on the observations $Y(0), Y(1), \ldots, Y(11)$ and the knowledge of $\overline{P}(0)$, we must first calculate

$$\hat{X}(0), \hat{X}(2), \ldots, \hat{X}(12)$$

along with their associated covariances

$$\overline{P}(1), \overline{P}(2), \ldots, \overline{P}(12)$$

from Eqs. (4.136)–(4.140). Note that the values of $\hat{X}(12), \hat{X}(11), \overline{P}(11)$ are needed for Eqs. (4.159) and (4.160) to obtain $\hat{X}^j(11-j)$ and $R(11-j)$. We shall start with $\hat{X}(12)$ in Eq. (4.159), and from Eq. (4.149) we shall obtain $\hat{X}^0(11)$. Thus, $\hat{X}^{(1)}(10)$, along with $R(10)$, can easily be obtained.

It is obvious that the filtering problem can be derived from the smoothing problem when $j = 0$.

4.13 NONLINEAR ESTIMATION

State estimation for nonlinear systems, whether the nonlinearity is introduced by the model generating the random process or by the observation function, is a formidable task. There are some techniques for solving some special problems, but, to date, no general closed form solution has been found. The most attractive alternate approach makes use of the concept of linearization, which we shall discuss for the discrete case only.

Suppose a physical system is expressed by

$$X(k+1) = f(X(k), k) + \overline{B}(k) U(k) \qquad (4.161)$$

$$Y(k) = g(X(k), k) + v(k) \qquad (4.162)$$

with statistics given by

$$EU(k) = 0 \qquad (4.163)$$

$$EU(k_1) U'(k_2) = \overline{Q}(k) \Delta(k_2 - k_1) \qquad (4.164)$$

$$Ev(k) = 0 \qquad (4.165)$$

$$Ev(k_1) v'(k_2) = \overline{L} \Delta(k_2 - k_1) \qquad (4.166)$$

where the noise terms $U(k)$ and $v(k)$ both appear in additive form.

To solve the estimation problem, we assume f and g are continuously differentiable. Thus we can expand them in Taylor series around a nominal solution $\widetilde{X}(k)$ and truncate after the second term (first order approximation). In many physical situations $X(k) - \widetilde{X}(k)$ can be characterized via a linear differential equation. Thus, the Kalman filtering algorithms can be used to obtain an estimate of $\widehat{X}(k)$ of $\widetilde{X}(k)$. Very often we shall set $\widetilde{X}(k) = \widehat{X}(k)$; see the appropriate references in Chapter 4. Therefore,

$$f(X(k), k) = f(\widehat{X}(k), k) + \frac{\partial f(\widehat{X}(k), k)}{\partial x} [X(k) - \widehat{X}(k)] \qquad (4.167)$$

$$g(X(k), k) = g(\widehat{X}(k), k) + \frac{\partial g(\widehat{X}(k), k)}{\partial x} [X(k) - \widehat{X}(k)] \qquad (4.168)$$

where $\partial f/\partial x$ and $\partial g/\partial x$ are given by

$$\frac{\partial f(\widehat{X}(k), k)}{\partial x} = \begin{bmatrix} \frac{\partial f_1}{\partial x_1} & \frac{\partial f_1}{\partial x_2} & \cdots & \frac{\partial f_1}{\partial x_n} \\ \frac{\partial f_2}{\partial x_1} & \frac{\partial f_2}{\partial x_2} & \cdots & \frac{\partial f_2}{\partial x_n} \\ \vdots & & & \\ \frac{\partial f_n}{\partial x_1} & \frac{\partial f_n}{\partial x_2} & \cdots & \frac{\partial f_n}{\partial x_n} \end{bmatrix}_{X=\widehat{X}} \qquad (4.169)$$

$$\frac{\partial g(\widehat{X}(k), k)}{\partial x} = \begin{bmatrix} \frac{\partial g_1}{\partial x_1} & \frac{\partial g_1}{\partial x_2} & \cdots & \frac{\partial g_1}{\partial x_n} \\ \frac{\partial g_2}{\partial x_1} & \frac{\partial g_2}{\partial x_2} & \cdots & \frac{\partial g_2}{\partial x_n} \\ \vdots & & & \\ \frac{\partial g_n}{\partial x_1} & \frac{\partial g_n}{\partial x_1} & & \frac{\partial g_n}{\partial x_n} \end{bmatrix}_{X=\widehat{X}} \quad (4.170)$$

The estimate $\widehat{X}(k)$ of $X(k)$ can be obtained by

$$\widehat{X}(k+1) = f(\widehat{X}(k), k) + F(k)[y(k) - g(\widehat{X}(k), k)] \quad (4.171)$$

where $F(k)$ is obtained by Eq. (4.138) and

$$\overline{A} \triangleq \frac{\partial f(\widehat{X}(k), k)}{\partial x} \quad (4.172)$$

$$\overline{C} \triangleq \frac{\partial g(\widehat{X}(k), k)}{\partial x} \quad (4.173)$$

The estimator represented by Eq. (4.171) is the one-step predictor. The linearization process for filtering and smoothing is similar.

The next example shows the application of estimation theory to a nonlinear problem. However, the continuity of the subsequent sections will not be affected if one skips over this example.

Example 13

This example describes a part of an estimation methodology that was developed for the Viking Orbiter High Gain Antenna (HGA) Pointing Control System. The Viking Orbiter was launched in 1975, and its telecommunication performance required that the HGA pointing error be held to 0.7 deg. with a confidence level of 99.7 %. A higher pointing error would have resulted in a degraded transmission rate. Without the use of the estimation methodology discussed here, the pointing error would have been exceeded by 43 %.

The antenna pointing error is the angle between the HGA boresight and an ideal earth-vector. This error is a function of many stochastic rotational errors. If we develop a model that yields corrections to these rotational errors, the result is a reduction of the HGA pointing error. The model discussed here used the signal strength measurements received at the ground tracking station as its observation. Using several successive observations yielded smaller pointing errors. The details will not be presented here for the sake of brevity, but can be found in ref. [35].

Let Q denote the spacecraft fixed body axis (X, Y, Z), Q' the spacecraft actual coordinate system, and Q_3 the antenna coordinate system (X_3, Y_3, Z_3) (see Figure 4-5). The other coordinate systems Q_1 and Q_2 designate intermediate coordinate systems (azimuth and elevation actuators). Let $\bar{e}_i$, for $i = 1,2,3,4$, denote the rotational error vectors in $Q, Q_1, Q_2,$ and Q_3, respectively.

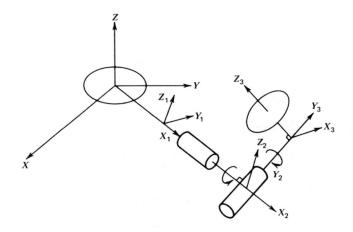

Fig. 4-5. Antenna Boresight Coordinate System

The scalar rotational errors are shown in Figure 4-6. If we denote the antenna error vector at $\bar{e}_A$, then it can be shown [35] that the first order antenna error $\bar{e}_A$ can be written as:

$$\bar{e}_A = \bar{e}_4 + T_{Q_3}^{Q_2} \bar{e}_3 + T_{Q_3}^{Q_1} \bar{e}_2 + T_{Q_3}^{Q_2} T_{Q_2}^{Q_1} T_{Q_1}^{Q'} \bar{e}_1$$

(4.175)

$$= \bar{e}_4 + T_{Q_3}^{Q_2} \bar{e}_3 + T_{Q_3}^{Q_1} \bar{e}_2 + T_{Q_3}^{Q'} \bar{e}_1$$

where T_{II}^{I} denotes the orthogonal transformation from space I to II. The signal strength measurement (X-band) at the ground station is modeled as

$$S = S_0 + 10 \log_{10} \left(\frac{\sin a\rho}{a\rho} \right)^2$$

(4.175)

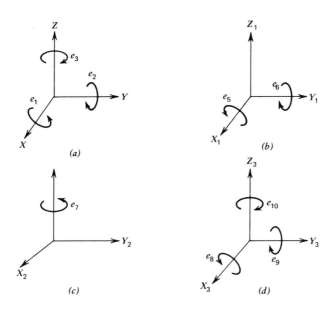

Fig. 4-6. Rotational Error Vectors

where S_0 is the antenna gain constant, a is constant, and ρ is the angle between the antenna boresight and the earth vector. The signal strength measurement is contaminated by many noise sources, such as atmospheric noise, electronic noise, etc.

Let the state $X(t)$ denote the vector

$$X(t) = [e_{10}, e_9, \ldots, e_1, \delta S_0, a] \qquad (4.176)$$

where δS is the signal strength measurement error from its peak S_0. The noise-corrupted observations are given by

$$Y(t) = S(X, t) + v(t) \qquad (4.177)$$

where $v(t)$ designates the lumped white noise source, and $S(X, t)$ designates the signal strength measurement at time t, and is dependent on X. From Eq. (4.175) it is obvious that Y is a nonlinear function of X. Since the rotational errors are assumed to be constant in-flight, then

$$\dot{X} = 0 \qquad (4.178)$$

Thus, from Eqs. (4.177)–(4.178), we have constructed a dynamic model for the Kalman filter.

Let X_0 be the nominal solution of X, and Y_0 be its corresponding observation. We can linearize the state equation (4.177) around X_0. Let W and Z denote $X - X_0$ and $Y - Y_0$, respectively, then $Z(t)$ can be approximated as:

$$Z(t) \cong C(t) W(t) + v(t) \tag{4.179}$$

where

$$C(t) = \frac{\partial S(X_0, t)}{\partial X} = \frac{\partial S(X_0, t)}{\partial \bar{\epsilon}_A} \frac{\partial \bar{\epsilon}_A}{\partial X} \tag{4.180}$$

Now the best linear m.s.e. of X can be obtained. The details are omitted. However, Figures 4-7 and 4-8 show the improvement of the uncertainties of the antenna rotational errors X_3 and Y_3 versus days past launch [35].

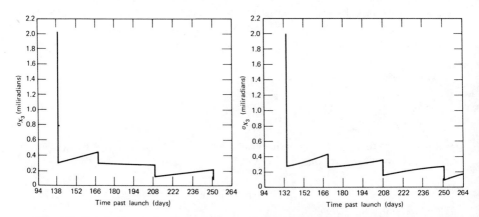

Fig. 4-7. X-Axis Uncertainties vs. Time   Fig. 4-8. Y-Axis Uncertainties vs. Time

4.14 REFORMULATION OF KALMAN FILTERING

The Kalman filtering algorithm is given by Eqs. (4.148)–(4.149). Another form of Kalman filter which is used in practical situations is given via:

$$\hat{X}^0(k) = \bar{P}^0(k) \{ [\bar{P}(k)]^{-1} \hat{X}(k) + \bar{C}'(k) [\bar{L}(k)]^{-1} Y(k) \} \tag{4.181}$$

and

$$\overline{P}^0(k) = \{[\overline{P}(k)]^{-1} + \overline{C}'(k)[\overline{L}(k)]^{-1}\overline{C}(k)\}^{-1} \qquad (4.182)$$

The proof follows by substituting $F(k)$ from Eq. (4.149 b) into $\overline{P}^0(k)$ given via Eq. (4.149 d) and using the following matrix identity which states: for any matrices M_1, M_2, and M_3 such that M_1, M_3, $(M_2 M_1^{-1} M_2' + M_3)$, and $(M_1 + M_2' M_3^{-1} M_2)$ are invertible, then

$$(M_1 + M_2' M_3^{-1} M_2)^{-1} = M_1^{-1} - M_1^{-1} M_2' (M_2 M_1^{-1} M_2' + M_3)^{-1} M_2 M_1^{-1}$$

(4.183)

If in the above equation, we change M_1 to $-M_1$, and use the new equation we can rewrite $F(k)$ as:

$$F(k) = \{[\overline{P}(k)]^{-1} + \overline{C}'(k)[\overline{L}(k)]^{-1}\overline{C}(k)\}\,\overline{C}'(k)[\overline{L}(k)]^{-1} \qquad (4.184)$$

The advantage of Eqs. (4.182) and (4.184) over the previous formulation is that if there is no information about $\overline{P}(k)$, then its inverse is zero. Also note that Eq. (4.149a) can be rewritten as

$$\hat{X}^0(k) = \hat{X}(k) + F(k)[Y(k) - \overline{C}(k)\hat{X}(k)] \qquad (4.185)$$

if we use Eq. (4.148).

4.15 DISCUSSION AND CONCLUDING REMARKS

There are two basic ideas, *controllability* and *observability*, that serve as the backbone of state estimation theory and optimal control. The concepts of controllability and observability are due to Kalman [36].

Basically, a state equation given by Eq. (4.83a) is said to be completely state controllable at time t_0, if for any state $X(t_0)$ and any other state X^1, we can obtain a finite time $t_1 > t_0$ and a control input over $[t_0, t_1]$ denoted by $U_{[t_0, t_1]}$ such that it will steer $X(t_0)$ to X^1 at time t_1. The state equation (4.83a) is said to be uncontrollable if it is not completely controllable.

The dynamic system given by Eq. (4.83) is completely observable at time t_0 if for any state $X(t_0)$, we can find a finite time $t_1 > t_0$, such that the information $U_{[t_0, t_1]}$ and its corresponding output $Y_{[t_0, t_1]}$ over the interval $[t_0, t_1]$ is sufficient to solve for $X(t_0)$ and consequently for all $X(t)$, where $t_0 \leq t \leq t_1$. A dynamic system is unobservable if it is not completely observable.

In estimation theory the concept of complete observability, that is, the knowledge of stochastic input $U(t)$ and its corresponding output $Y(t)$, allows us to solve for the state $X(t)$. Note that the optimal solution is given via the minimization of the state equation; see references [5] and [6] for more detail.

EXERCISES

4.1 Given $X_1, X_2, \ldots, X_n$ as random variables such that:

$$E(X_i) = m \text{ and } \text{var}(X_i) = \sigma^2$$

Assume that $X_i - m$ and $X_j - m$ are orthogonal for $i \neq j$. Let

$$\hat{m} = \frac{1}{n} \sum_{i=1}^{n} X_i$$

and

$$\hat{\sigma}^2 = \frac{1}{n-1} \sum_{i=1}^{n} (X_i - \hat{m})^2$$

be estimates of m and σ.

(a) Determine whether or not $\hat{m}$ is unbiased.

(b) Show that

$$\left[\sum_{i=1}^{n} (X_i - m) \right]^2 = \sum_{i=1}^{n} (X_i - m)^2 + \sum_{i \neq j}^{n} \sum (X_i - m)(X_j - m)$$

Hint: First prove that

$$n(\hat{m} - m) = \sum_{i=1}^{n} (X_i - m)$$

(c) Determine whether or not $\hat{\sigma}^2$ is an unbiased estimate of σ^2.

4.2 Let the random variables X_1 and X_2 be such that:

$$E(X_1) = E(X_2) = m$$

and

$$\text{var}(X_1) = \text{var}(X_2) = \sigma^2$$

with

$$E[(X_1 - m)(X_2 - m)] = 0$$

(a) If $\mathbf{X} = (X_1, X_2)$, then show that:

$$C_{m\mathbf{X}} = (m^2, m^2)$$

(b) Show that $E(X_1^2) = E(X_2^2) = \sigma^2 + m^2$ and $E(X_1, X_2) = m^2$.

(c) Obtain the covariance of $\mathbf{X}$.

(d) Obtain the m.s.e. of m from the data (x_1, x_2).

(e) Determine the conditions such that your m.s.e. in part (d) is unbiased.

4.3 Let $R(\tau)$ be the autocorrelation function of a process $X(t)$. Suppose it is desired to obtain the linear m.s.e. of $X(t + \lambda)$ for some $\lambda > 0$ in terms of $X(t)$, $X'(t)$, and $X''(t)$ i.e., $\hat{X}(t + \lambda) = a_1 X(t) + a_2 X'(t) + a_3 X''(t)$. Use the orthogonality principle to determine the optimum estimate of $\hat{X}(t + \lambda)$ and determine the m.s.e. of the error $X(t + \lambda) - \hat{X}(t + \lambda)$.

4.4 The zero mean random variable X is to be estimated in the linear mean square sense by the random variables $Y_1, Y_2, \ldots, Y_n$ each of mean zero. Let $\hat{X}$ be such an estimate. Utilizing the orthogonality principle:

(a) Show that $E(e^2) = E[(X - \hat{X})^2] = E[(X - \hat{X})X]$.

(b) Obtain the optimal solution $\hat{X}$.

(c) If e_m is the error corresponding to the optimal solution, i.e., $e_m = X - \hat{X}$, then verify whether or not

$$E(e_m^2) = \frac{\det\left\{E\begin{bmatrix} X \\ Y_1 \\ \cdot \\ \cdot \\ \cdot \\ Y_n \end{bmatrix} [X\ Y_1\ \cdots\ Y_n]\right\}}{\det\left\{E\begin{bmatrix} Y_1 \\ \cdot \\ \cdot \\ \cdot \\ Y_n \end{bmatrix} [Y_1\ \cdots\ Y_n]\right\}}$$

4.5 Let $Y(t) = s(t) + n(t)$ be given such that $En(t) = Es(t) = 0$.

(a) Use the orthogonality principle to estimate $\dot{s}(t) = (d/dt) s(t)$, and show that the optimal estimate (unrealizable) $\hat{\dot{s}}$ can be obtained from:

$$R_{\dot{s}Y}(\tau) = \int_{-\infty}^{\infty} R_Y(\tau - \lambda) \hat{h}(\lambda) \, d\lambda$$

where $\hat{h}(\lambda)$ is the optimum impulse response.

(b) Show that

$$\hat{H}(j\omega) = \frac{j\omega S_{sY}(\omega)}{S_Y(\omega)}$$

Hint: $S_{\dot{s}Y}(\omega) = j\omega S_{sY}(\omega)$.

(c) Given $R_s(\tau) = \exp(-|\tau|)$ or $S_s(\omega) = 2/(1 + \omega^2)$ and $R_n(\tau) = 2\delta(\tau)$, obtain an optimum estimate $\hat{\dot{s}}$ with the constraint of realizability imposed.

(d) In part (c) design an optimum realizable predictor $\hat{\dot{s}}(t + 1)$.

(e) Design an optimum realizable filter for

$$W(t) = \int_0^{\infty} s(t - \lambda) \, d\lambda$$

The answers in parts (c)–(e) can be left in the frequency domain.

4.6 A model is generated when white noise with the variance of unity (unity spectral density) is passed through a system with the transfer function $1/[s(s + 1)]$. The model is also contaminated with white noise $n(t)$ with $S_n(\omega) = 1$. Assume that $E(s(t) n(t)) = 0$. Find the transfer function $\hat{H}(s)$ of optimum estimate that will yield the best m.s.e. Also obtain the transfer function of the best m.s.e. of the derivative.

4.7 Consider the RC network given by:

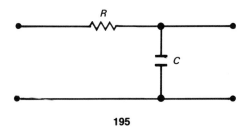

where the unit impulse response $h(t)$ is given by:

$$h(t) = \frac{1}{\alpha}\exp(-t/\alpha), \text{ with } \alpha = RC$$

Let the input to the filter be $y(t)$ given by:

$$Y(t) = s(t) + n(t)$$

where $s(t)$ is given by:

$$s(t) = A\cos(\omega_0 t + \theta) \text{ volts}, \quad \omega_0 = \frac{2\pi}{T}$$

with the random variable θ distributed uniformly over $[0, 2\pi]$. The amplitude A is constant, and $n(t)$ is a zero mean white noise with its power spectrum given by

$$S_n(\omega) = N \text{ (watts/Hz)}$$

(a) Calculate the input power spectrum.
(b) Calculate the input power.
(c) Calculate the output power due to the signal only.
(d) Calculate the output power due to the noise only.
(e) If the signal-to-noise ratio (SNR) is given by:

$$\text{SNR} = \frac{\text{Output power due to signal}}{\text{Output power due to noise}}$$

then obtain the maximum SNR.

4.8 Let $Y(t)$ be an observation given by:

$$Y(t) = s(t) + n(t)$$

where

$$S_s(\omega) = \frac{1}{\omega^4}, \; S_n(\omega) = 4, \; \text{and} \; S_{ns}(\omega) = 0$$

(a) Find the optimum predictor $\hat{s}(t + \lambda)$ by finding the corresponding optimum impulse response without the constraint of physical realizability.

(b) Repeat part (a) with the constraint of realizability imposed.
Hint: you may need to use

$$1 + k^2 s^4 = (1 + \sqrt{2k}\, s + ks^2)(1 - \sqrt{2k}\, s + ks^2)$$

You may leave your answers in the frequency domain.

4.9 Let X be a scalar random variable and $\hat{X}_1$ and $\hat{X}_2$ be two correlated unbiased estimates of X with associated variances (covariances) σ_1^2 and σ_2^2, respectively. Let ρ denote:

$$\rho = E[(X - \hat{X}_1)(X - \hat{X}_2)]$$

and σ^2 denote the variance (covariance) associated with X, where $\hat{X} = \alpha \hat{X}_1 + \beta \hat{X}_2$.

(a) Show that $\alpha + \beta = 1$ and derive an expression for σ^2 in terms of σ_1^2, σ_2^2, ρ, α, and β.

(b) Obtain the optimal estimate $\hat{X}$, i.e., determine α and β such that $\hat{X}$ is optimal.

4.10 Let a system be described via the model:

$$\begin{bmatrix} \dot{X}_1 \\ \dot{X}_2 \end{bmatrix} = \begin{bmatrix} X_1(t) \\ X_2(t) \end{bmatrix} + \begin{bmatrix} U_1(t) \\ U_2(t) \end{bmatrix}$$

and

$$Y = \begin{bmatrix} X_1 \\ X_2 \end{bmatrix} + \begin{bmatrix} v_1 \\ v_2 \end{bmatrix}$$

where

$$E[U\,U'] = E[v\,v'] = I\,\delta(t-\tau)$$

Note that U and v are vectors. Write the appropriate equations for the optimal estimate. What is the error covariance matrix?

4.11 Suppose it is desired to estimate a constant which is unknown; a system model may be given by:

$$\dot{X} = 0, \quad Y = X + v$$

where

$$E[v(t)\,v(\tau)] = Q\,\delta(t-\tau)$$

Obtain a closed form optimal solution.

4.12 Repeat problem 4.11 if the state model is changed to:

$$\dot{X} = \frac{-1}{2}X + U(t)$$

and $Q = 1/4$, $E[U(t)\,U(\tau)] = 2\,\delta(t-\tau)$, and $E[Uv] = 0$.

4.13 A scalar discrete random process $X(k)$ is given by:

$$X(k+1) = 0.5\,X(k) + U(k)$$

$$Y(k) = X(k) + v(k)$$

where $U(k)$ and $v(k)$ are white noise terms such that:

$$E[v^2(k)] = E[U^2(k)] = 1$$

Also assume that:

$$EX(0) = 0$$

$$E[X(0)]^2 = 1$$

It is obvious that the Kalman estimator (one-step predictor) is given by:

$$\hat{X}(k+1) = [0.5 - F(k)] \hat{X}(k) + F(k) Y(k)$$

$$F(k) = \frac{0.5 P(k)}{P(k) + 1}$$

$$P(k+1) = [0.5 - F(k)]^2 P(k) + 1 + F^2(k)$$

$$P(0) = 1, \hat{X}(0) = 0$$

Suppose $Y(2)$ is not received, then perform the following:

(a) Provide the correction (or the adjustment) necessary in the above Kalman estimator to account for $Y(2)$ not being received.

(b) Calculate the loss in terms of estimation error variance associated with $\hat{X}(3)$ in part (a). The error variance is denoted by $E(3)$ and is given by:

$$E(3) = \overline{P}(3) - P(3)$$

where $\overline{P}(3)$ is the covariance with the observation $Y(2)$ missing.

(c) Calculate the steady-state covariance

$$P_{ss} = \lim_{k \to \infty} P(k)$$

4.14 Let $s(t)$ be a signal such that

$$s(t) = \begin{cases} 2, & t \in [0,1] \\ 0, & \text{otherwise} \end{cases}$$

and the observation given by

$$y(t) = s(t) + n(t)$$

where $n(t)$ is a white noise. Obtain the filter that yields the maximum signal-to-noise ratio.

4.15 Let a signal $s(t)$ be given

$$s(t) = \begin{cases} 1 + t, & 0 \leq t \leq T \\ 0, & \text{otherwise} \end{cases}$$

and the observation by

$$y(t) = s(t) + n(t)$$

where $n(t)$ is a white noise independent process. Derive a linear filter represented by $H(j\omega)$ such that the signal-to-noise ratio is maximized at time T.

4.16 If X and Y are jointly normal, show that $\hat{X}$ coincides with $E(X|Y)$.

4.17 Let Y and Z be uncorrelated random vectors, then the optimal solution of X in terms of Y and Z is given by:

$$\hat{X}(W) = \hat{X}(Y) + \hat{X}(Z)$$

where $W = [Y \vdots Z]'$.

4.18 In the above problem, show that

$$\hat{X}(W) = \hat{X}(Y) + \hat{X}(\tilde{Z})$$

where $\tilde{Z} = Z - \hat{Z}(Y)$.

CHAPTER 5
APPLICATION OF ESTIMATION THEORY TO IMAGE RESTORATION

5.1 INTRODUCTION

This chapter illustrates the application of estimation theory to image restoration. In general, image restoration is an operation performed on a degraded image in order to improve its quality. In theory, restoring an image by use of classical filtering techniques does not seem to be difficult, since the processing of signals in one dimension can be extended to the two-dimensional case by means of linear system theory. However, the application is cumbersome and may become impractical when a large amount of data is to be processed. Every classical technique has a drawback because it is nonrecursive and is seriously hampered by the presence of noise and other disturbances.

The image distortion is due to various sources such as reflections from spurious objects, aberration, atmospheric turbulence, and degradation introduced during transmission. A block diagram representing the image restoration problem is shown in Figure 5-1. The undistorted image is represented by the real function $b(x, y)$, where $b_d(x, y)$ and $\hat{b}(x, y)$ represent the degraded image and its estimate (restored image), respectively. In the actual case, a distorted image is observed, and the true image is to be recovered. In many cases, the observed image can be adequately modeled as an ideal image corrupted by additive noise terms. Thus, the image and the noise sources may be assumed to be sample functions of random processes with known statistics (first two moments). The noise sources are lumped together as a single additive noise term.

This chapter is concerned with the solution of the statistical image restoration problem. The image restoration problem as described above becomes the classi-

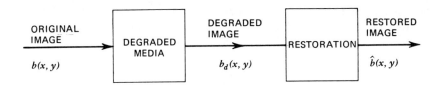

Fig. 5-1. Image Restoration Problem

cal problem of statistical estimation and filtering, where we attempt to filter out the noise from the observation. The most general and computationally efficient procedure is via Kalman filtering and its nonlinear extension. To apply this technique, the observation must be a function of one independent variable, in contrast to the brightness $b(x, y)$ or $b_d(x, y)$, which are functions of two spatial random variables, where

$$b_d(x, y) = b(x, y) + v(x, y)$$

and $v(x, y)$ denotes the spatial noise term.

Since the noise is additive, we can convert the spatial brightness $b(x, y)$ into one dimension, without any loss of generality, as the output of an optical scanner. For the noise term $v(x, y)$, the procedure would be similar. If we scan the image horizontally with an optical scanner moving at uniform speed, the scanner output denoted by $s(t)$, where t is the time variable, will be a one-dimensional process. The required steps taken for the derivation to enhance the image (via Kalman filtering) are given below.

(1) Obtain $s(t)$ via the optical scanner as described above.

(2) Determine the autocorrelation function of $s(t)$.

(3) Determine the dynamical state-variable model for $s(t)$.

(4) Repeat step 3 with the noise term.

(5) Design an appropriate recursive estimator utilizing the above dynamical state-variable model.

The characterization of a two-dimensional signal (image) via a recursive estimator was first developed in reference [1] and further extended in references [2], [3], and [4]. In all the references a simple form of two-dimensional signal was assumed. However, the one dimensional signal $s(t)$ became nonstationary and nonseparable, references, [5]-[9], because of the scanner's periodic movement. It is shown that for such processes, no suitable dynamical model representing the statistics of $s(t)$ suitable for Kalman filtering exists.

To remedy the dynamical modeling problem which is caused by the nonstationarity of $s(t)$ we proceed by generating another stochastic process whose autocorrelation function is wide-sense stationary and which approximates the autocorrelation function of $s(t)$. As will be seen later, the results obtained using this technique are very satisfactory. Since we shall be dealing with the realization of autocorrelation functions by a state-variable dynamical model (spectral factorization), a brief background of spectral factorization is given.

Next, we shall utilize a better approximation to $s(t)$ (scanner's output) or its autocorrelation function developed by partitioning the image into a collection of vertical strips and approximating $s(t)$ by a series of stationary random processes, one associated with each strip. For each stationary approximation, a corresponding linear time-invariant dynamic model is constructed. A procedure for recursively enhancing a degraded image is developed in a manner similar to the case where the image has not been partitioned. The major difference is that rather than utilizing one dynamical model corresponding to one autocorrelation function, a chain of dynamic models corresponding to many autocorrelation functions is considered. Examples are constructed to show the effectiveness of this enhancement process.

5.2 SPECTRAL FACTORIZATION

The concept of spectral factorization has become increasingly more important since Wiener's original work [8] on the subject. Basically, spectral factorization determines the equations that describe a linear system when the system is driven by white noise and the covariance of the output is known. Whenever the covariance function of a process is driven by white noise via a system of differential equations of first order, we refer to this system as a dynamical model. More specifically, given a covariance function $R(t, \tau)$, where $t \leq t_1$ and $\tau \leq \tau_1$ for some fixed t_1 and τ_1, the factorization problem is to determine a realizable linear filter (differential equation model) that, when driven by white noise, yields $R(t, \tau)$ as its output covariance.

It is well known, [6] and [10], that, in general, no such realization may exist. However, if its existence were guaranteed, the representation (in some sense) would be unique. In its most popular form, the spectral factorization would be confined to stationary situations. Then the corresponding dynamical model under consideration would be time-invariant, and the white noise forcing function must have started infinitely in the past. This dynamical model would be asymptotically stable. It is also desirable to deal with finite-dimensional dynamical models, implying that each linear model must possess a rational bilateral Laplace transform. We can summarize the above discussion by the statement of Theorem 1, which we shall not prove, but which is proved in reference [12].

Theorem 1

A necessary and a sufficient condition that a stationary process $y(t)$ be representable as the output of an asymptotically stable, time-invariant, finite, dimensional linear model is that is spectral density $S(s)$ be a rational function of the form $H(s)H(-s)$, with

$$H(s) = \frac{M(s)}{\rho(s)} \tag{5.1}$$

for some polynomial

$$\rho(s) = s^n + \sum_{i=0}^{n-1} \alpha_i s^i$$

with all roots in the left half part of the s-plane and

$$M(s) = \sum_{i=0}^{n-1} \beta_i s^i$$

with degree less than or equal to $n - 1$ and all roots in the left half of the s-plane, where α_i and β_i are the real coefficients. That is, $H(s)$ has all of its poles and zeros in the left half of the s-plane.

5.2.1 Determination of the Output Covariance From a Linear Dynamical Model

Consider the following dynamical model, given by:

$$\dot{x} = A(t) x(t) + B(t) u(t)$$

$$y(t) = C(t) x(t) \tag{5.2}$$

where $x(t)$ is an $n \times 1$ vector, u is an $m \times 1$ vector, y is a scalar, A, B, and C are matrices of appropriate dimensions (not necessarily time-invariant), and $u(t)$ is a zero-mean white noise vector, such that:

$$Eu(t)u'(\tau) = K\delta(t - \tau) \tag{5.3}$$

where K is an $m \times m$ symmetrical matrix and prime denotes the transpose.

It is desired to calculate the output covariance (an autocorrelation, since $y(t)$ is of zero mean) $Ey(t)y(\tau)$, given by:

$$Ey(t)y(\tau) = C(t) Ex(t) x'(\tau) C'(\tau) \tag{5.4}$$

Let the random variable $x(t_0)$, where t_0 is the initial time, be statistically independent of $u(t)$. It is well known that the solution of $x(t)$ is given by:

$$x(t) = \Phi(t, t_0) x(t_0) + \int_{t_0}^{t} \Phi(t, \tau) B(\tau) u(\tau) d\tau \tag{5.5}$$

where $\Phi(t, \tau)$ is the state transition matrix; i.e.,

$$\frac{d\Phi(t, \tau)}{dt} = A(t) \Phi(t, \tau) \tag{5.6}$$

$$\Phi(t, t) = I \tag{5.7}$$

Substituting $x(t)$ from Eq. (5.5) into (5.4) and performing some mathematical operations, we obtain [20]:

$$Ey(t)y(\tau) = C(t) \Phi(t, \tau) P_x(\tau) C'(\tau) 1(t - \tau) + C(t) P_x(t) \Phi'(t, \tau) C'(t) 1(\tau - t) \tag{5.8}$$

$$P_x(t) = Ex(t) x'(t) \tag{5.9}$$

where $1(t)$ denotes the unit step function.

From the dynamical model (Eq. 5.2), $P_x(t)$ can be shown to be the solution of the differential equation [10]:

$$\dot{P}_x = AP_x + P_x A' + BKB' \tag{5.10}$$

where the covariance $P_x(t_0)$ must be given.

5.2.2 Independence of Estimation Problem of a Particular Coordinate System

In spectral realization, $y(t)$, given by Eq. (5.2), is the signal without any noise contamination. Often, we receive a contaminated observation $z(t)$, given by:

$$z(t) = y(t) + n(t) \tag{5.11}$$

where $n(t)$ is additive noise, which is assumed to be uncorrelated with $y(t)$. In [10] it is shown that the only information necessary for recursive estimation is the knowledge of $Ey(t)\,y(t+\tau)$ and $Ez(t)\,z(t+\tau)$. That is, the solution of recursive estimation in the mean-square sense is independent of the particular coordinate system for model $z(\cdot)$ and $y(\cdot)$ processes; hence, a unique solution associated with minimum mean-square estimation can be obtained where the models for the processes are not given in advance. All these models are related to one another by a linear transformation. For example, if

$$\dot{x} = Ax(t) + Bu(t)$$

$$y = Cx(t) + v(t) \tag{5.12}$$

and

$$\dot{x}^* = A^* x^*(t) + B^* u^*(t)$$

$$y = C^* x^*(t) + v(t) \tag{5.13}$$

correspond to the same realization, then there exists a linear transformation $T(t)$ such that:

$$x^*(t) = T(t)\,x(t) \tag{5.14}$$

and

$$\hat{x}^* = T(t)\hat{x}(t) \tag{5.15}$$

where $\hat{x}$ and $\hat{x}^*$ are the estimates corresponding to Eqs. (5.12) and (5.13), respectively.

The covariance estimates can be obtained accordingly.

5.3 RECURSIVE IMAGE ESTIMATION

5.3.1 Procedure Outline

The enhancement of images that are characterized only by statistical data where the picture contains additive noise is considered in this section. The random process representing the output scanner is characterized by the output of a dynamical model with white noise input. The dynamical model describes the first-order vector Markov process. The procedure of Kalman filtering is then utilized to recursively determine the minimum mean-square error estimate of the image. The result is also extended to obtain the smoothing of data. Two examples, one with very high SNR, are used to illustrate the effectiveness of the procedure. In what follows, the image is assumed to be a two-dimensional, stationary correlation function of zero mean. Thus, the autocorrelation function and the covariance become identical. The statistical information about the image and the noise is assumed to be known and uncorrelated, and the noise is additive.

5.3.2 Derivation of Autocorrelation Function of Scanner Output

Let us scan a picture horizontally using an optical scanner denoted by $s(t)$. Let the horizontal position (a continuous variable) be denoted by z, where $0 \leqslant z \leqslant Z$, and the vertical variable by an integer $n = 1, 2, \cdots, N$ representing the nth scanned line. The brightness function is defined by $b(z, n)$. Let us assume, without any loss of generality that $b(z, n)$ is of zero mean. The random process $b(z, n)$ is assumed to be wide-sense stationary, with the autocorrelation function defined by:

$$Eb(z_2, n_2) b(z_1, n_1) \doteq R(z_2 - z_1, n_2 - n_1) \doteq R(z, n) \tag{5.16}$$

Assume that the scanner output $s(t)$ has a horizontal speed $v = 1$ and, without any loss of generality, that the vertical movement takes zero time.

Let us determine $Es(t)\,s(t+\tau)$ in terms of $R(z,n)$ and Z. The variables t and τ can be equivalently expressed by:

$$\begin{cases} t = jT + \sigma, & j = 0,1,2,\cdots,N-1, \quad 0 \leqslant \sigma \leqslant T \\ \tau = iT + \gamma, & i = \cdots,-1,0,1,\cdots \\ 0 \leqslant t + \tau \leqslant NT, & \quad 0 \leqslant \gamma \leqslant T \end{cases}$$

(5.17)

where $T = Z$ is the time required to traverse one horizontal line. The scanner output can now be written as:

$$s(t) = b(\sigma, j+1),\ s(t+\tau) = \begin{cases} b(\sigma + \gamma, i + j + 1), & \text{if } \sigma + \gamma \leqslant T \\ b(\sigma + \gamma - T, i + j + 2), & \text{if } \sigma + \gamma > T \end{cases}$$

(5.18)

Now, utilizing Eqs. (5.16) and (5.17), we can obtain:

$$Es(t)\,s(t+\tau) = \begin{cases} R(\gamma, i), & \text{if } \sigma + \gamma \leqslant T \\ R(\sigma + \gamma - T, i + j + 2), & \text{if } \sigma + \gamma > T \end{cases}$$

(5.19)

It is clear that $Es(t)\,s(t+\tau)$ is a function of both σ and γ, or, equivalently, of t and τ; thus, it must be nonstationary. The nonstationarity is due to the edge condition. A simple check shows that $Es(t)\,s(t+\tau)$ is also periodic and a nonseparable function. It can be demonstrated that no finite-dimensional linear realization of this nonseparable autocorrelation exists.

We shall now seek to generate a random process denoted as $q(t)$ such that it has a stationary autocorrelation function which approximates the autocorrelation of the process $s(t)$. To generate $q(t)$, we proceed as follows. For a given t, $q(t)$ is defined by:

$$q(t) = s(jT + \xi) \qquad (5.20)$$

where ξ is assumed to be uniformly distributed over $[0, T]$. We shall now prove the following theorem.

Theorem 2

The random process $q(t)$ defined by Eq. (5.20) is stationary.

Proof

It is easy to verify that:

$$Eq(t) = 0$$

by the construction of $q(t)$.

Next, we must prove that $Eq(t)\,q(t + \tau)$ is a function of τ (or, equivalently, γ) only. To accomplish this end, we calculate the correlation function of the process $q(t)$:

$$Eq(t)\,q(t + \tau) = E_\xi E_s\,[s(jT + \xi)\,s(jT + \xi + iT + \gamma)]$$

$$= \frac{1}{T} \int_0^T E_s[s(jT + \xi)\,s(jT + \sigma + iT + \gamma)]\,d\sigma$$

$$(5.21)$$

This equation is obtained by utilizing Eq. (5.24) and $\tau = iT + \gamma$, which is given by Eq. (5.17) and the fact that ξ is uniformly distributed over $[0, T]$.

The subscripts s and ξ in (5.21) denote the expectation with respect to s and ξ, respectively. From Eqs. (5.19) and (5.21), one obtains:

$$Eq(t)\,q(t+\tau) = \frac{1}{T}\left[\int_0^{T-\gamma} R(\gamma, i)\,d\xi \right.$$

$$\left. + \int_{T-\gamma}^{T} R(T-\gamma, i+1)\,d\xi \right]$$

$$= \frac{T-\gamma}{T} R(\gamma, i) + \frac{\gamma}{T} R(T-\gamma, i+1) = r(\tau)$$

(5.22)

where $Eq(t)\,q(t+\tau)$ is defined as $r(\tau)$, which is a function of τ (or γ) only.

It is interesting to note that the correlation function of $q(t)$, namely, $r(\tau)$, can also be obtained by averaging the autocorrelation function of $s(t)$ over one period. However, it is important to mention that such averaging over the subintervals of a period may not give rise to a stationary autocorrelation function, and, furthermore, may not yield an autocorrelation function at all.

As an example, consider a scalar random process characterized by a scalar differential equation:

$$\dot{x} = -x + u$$

$$y(t) = \cos(t)\,x(t)$$

where the initial state $x(0) = 1/2$ and

$$Eu(t) = 0$$

$$Eu(t_1)\,u(t_2) = \delta(t_2 - t_1)$$

Then, the autocorrelation of $x(t)$ can be obtained as follows:

$$Ex(t)\,x(t+\tau) = \frac{1}{2}\exp(-|\tau|)$$

Thus, $Ey(t)\,y(t+\tau)$ is given by:

$$Ey(t)\,y(t+\tau) = \frac{1}{2}\cos(t)\cos(t+\tau)\exp(-|\tau|)$$

which is clearly nonstationary, since the correlation function of $y(t)$ depends on both t and $t+\tau$ and is periodic (of periodicity 2π). However, if we averaged this autocorrelation over $[0,\pi/4]$, the resulting average would depend on both t and $t+\tau$.

The randomization of ξ over the period T has the intuitive appeal that all points of the picture are weighted equally.

The following salient properties of $r(\tau)$ will be used in what follows:

$$r(iT) = R(0, i) \tag{5.23}$$

Since $R(z, n)$ is an autocorrelation function,

$$R(0, n) \geq R(z, n) \tag{5.24}$$

Thus, from (5.22) and (5.23),

$$\frac{r(iT + \gamma)}{r(iT)} \leq 1, \quad \text{for all } i, \gamma \tag{5.25}$$

The above properties indicate that, in general, the correlation function $r(\tau)$ has a periodic nature.

Example 1

Consider a square picture subdivided into a 32 × 32 grid. Let $T = 1$ second and $v = 1$. The signal is a 12 × 12 square starting at the 13th row and 13th column. Let m and n represent specific rows and columns, respectively. The above signal is represented by the brightness level $b(m, n) = 6.1$ where the signal exists and −1 otherwise, resulting in a zero mean sample function. As a first approximation, let us choose:

$$R(x, i) = \alpha \exp(-\mu_h |z| - \mu_v |i|)$$

where α, μ_h, and μ_v are to be determined. Computation of the sample power results in $\alpha = R(0,0) \approx 6.1$. The correlation between two adjacent grid points is calculated as 5.33, which is a value for $R(1/32,0)$ or $R(0,1)$. Hence,

$$R(x, i) = 6.1 \exp(-4.35|z| - 0.136|i|)$$

The correlation function is obtained by substituting the above into Eq. (5.22), and the plot is shown in Figure 5-2.

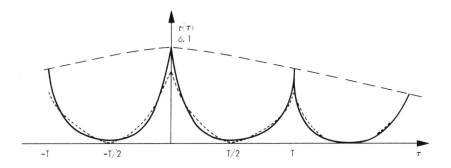

Fig. 5-2. Plot of r(τ) and r_a(τ) (Dashed Curve) as a Function of τ

5.3.3 Dynamical Modeling of Image Statistics

In this section, we wish to derive a differential equation model whose solution has an autocorrelation function approximating $r(\tau)$ given by Eq. (5.22). Since we subsequently intend to utilize a Kalman estimator, we seek a dynamical model of the form:

$$\dot{x}(t) = Ax(t) + Bu(t)$$

$$y(t) = Cx(t) \qquad (5.26)$$

where $x(t)$ is an n-dimensional vector, $u(t)$ is a white noise vector, and $y(t)$ is the scalar signal whose autocorrelation function is $r(\tau)$.

The procedure followed is to represent an approximation to $r(\tau)$, denoted by $r_a(\tau)$, as a sum of terms such that each term can be easily modeled, since, in general, $r(\tau)$ may not have a rational bilateral transform. The properties of $r(\tau)$ may be utilized to decompose $r(\tau)$ into the product of two functions $h(\tau)$ and $r(\tau)/h(\tau)$:

$$r(\tau) = \frac{r(\tau)}{h(\tau)} h(\tau) \qquad (5.27)$$

where $h(\tau)$ is chosen to satisfy:

$$h(iT) = R(0, i), \quad \text{for all } i \qquad (5.28)$$

Since in many practical cases the two-dimensional correlation function $R(z, i)$ is a monotonically decreasing function of i, a natural candidate for $h(\tau)$ is, in those instances, a combination of negative exponentials; i.e.,

$$h(\tau) = \sum_{i=1}^{I} l_i \exp(-\lambda_i |\tau|) \qquad (5.29)$$

The function $p(\tau)$ is then chosen to be a periodic function approximating $r(\tau)/h(\tau)$. The approximate correlation function is:

$$r_a(\tau) = h(\tau) p(\tau) \qquad (5.30)$$

Utilizing Eqs. (5.23) and (5.28), it can be seen that the function $r(\tau)/h(\tau)$ is unity at iT and less than unity for all other τ; furthermore, from (5.22) and (5.29), it is an even function. Hence, $p(\tau)$ is chosen to be an even function with period T. Thus, a natural candidate for this function is:

$$p(\tau) = \sum_{j=0}^{J} a_j \cos \frac{2\pi j}{T} \tau \qquad (5.31)$$

Consequently, an element of the function $r_a(\tau)$ has the form:

$$l_i \, a_i \exp(-\lambda_i |\tau|) \cos \frac{2\pi j}{T} \tau \qquad (5.32)$$

and there are $(J + 1) I$ such elements.

A differential equation model with white noise input can be simply constructed [8] to model each of these terms. Each will be a second-order system except for those corresponding to $j = 0$; i.e.,

$$l_i \, a_0 \exp(-\lambda_i |\tau|)$$

which will be of first order. If the white noise forcing functions (one being necessary for each i, j pair) are chosen to be mutually independent, the collection of all these differential equations defines the parameters A, B, C and represents the desired model for $r_a(\tau)$.

In the course of selecting the approximate function $r_a(\tau)$, we must choose the coefficients properly, such that $r_a(\tau)$ is a correlation function. We shall either guarantee that $r_a(\cdot)$ is a positive definite function or, equivalently, that the spectral density of $r_a(\tau)$ is positive [14].

Example 2

Using Example 1, let us derive a dynamic model for $r(\tau)$. Assume that the desired model has the form given by Eq. (5.26), and further that:

$$Eu(t) \, u(t + \tau)' = K\delta(\tau) \qquad (5.33)$$

where $\delta(\tau)$ is the Dirac delta function, the prime denotes the transpose, K is a positive definite matrix, and

$$Ey(t) \, y(t + \tau) = r_a(\tau) \qquad (5.34)$$

Because of the exponential nature of $R(z, i)$, we choose:

$$h(\tau) = R(0,0) \exp(-0.136|\tau|) \qquad (5.35)$$

and

$$p(\tau) = \sum_{j=0}^{J} a_j \cos 2\pi j\tau \qquad (5.36)$$

In this example, we use the notation μ_ν instead of 0.136.

The modeling procedure can be broken down as follows. The first term $r_a(\tau)$, namely,

$$a_0 \exp(-\mu_\nu |\tau|)$$

has the bilateral transform:

$$\frac{2\mu_\nu a_0}{(s+\mu_\nu)(s-\mu_\nu)} \doteq R_1(s) \qquad (5.37)$$

The function $R_1(s)$ can now be factored into two functions, $H_1(s)$ and $H_1(-s)$, where

$$R_1(s) = \frac{\sqrt{2a_0\mu_\nu}}{(s+\mu_\nu)} \frac{\sqrt{2a_0\mu_\nu}}{(s-\mu_\nu)}$$

and

$$H_1(s) = \frac{\sqrt{2a_0\mu_\nu}}{s+\mu_\nu}$$

Utilizing the method of this section, a dynamic realization of $H_1(s)$ is obtained as:

$$\dot{x}_1 = -\mu_\nu x_0(t) + \sqrt{2a_0\mu_\nu}\, u_1(t)$$

$$y_1(t) = x_1(t) \qquad (5.38)$$

The second term of $r_a(\tau)$, namely,

$$a_1 \exp(-\mu_\nu |\tau|) \cos 2\pi\tau$$

has the following bilateral transform:

$$R_2(s) = \frac{2a_1\mu_\nu[-s^2 + (2\pi)^2 + \mu_\nu^2]}{[(s+\mu_\nu)^2 + (2\pi)^2][(-s+\mu_\nu)^2 + (2\pi)^2]}$$

The function $R_2(s)$ can be factored out into two functions, $H_2(s)$ and $H_2(-s)$:

$$R_2(s) = \frac{\sqrt{2a_1\mu_\nu}[s + \sqrt{(2\pi)^2 + \mu_\nu^2}]}{(s+\mu_\nu)^2 + (2\pi)^2}$$

$$\times \frac{\sqrt{2a_1\mu_\nu}[-s + \sqrt{(2\pi)^2 + \mu_\nu^2}]}{(-s+\mu_\nu)^2 + (2\pi)^2}$$

where $H_2(s)$ is given by:

$$H_2(s) = \frac{\sqrt{2a_1\mu_\nu}[s + \sqrt{(2\pi)^2 + \mu_\nu^2}]}{(s+\mu_\nu)^2 + (2\pi)^2}$$

The corresponding dynamic realization of $H_2(s)$ is given as:

$$\dot{x}^{(2)} = A^{(2)}x^{(2)}(t) + B^{(2)}u^{(2)}(t)$$

$$y^{(2)}(t) = C^{(2)}x^{(2)}(t)$$

where the superscript denotes the model corresponding to the appropriate term. The coefficients $A^{(2)}$, $B^{(2)}$, and $C^{(2)}$ are given as:

$$A^{(2)} = \begin{bmatrix} 0 & 1 \\ -(2\pi)^2 + \mu_\nu^2 & -2\mu_\nu \end{bmatrix}$$

$$B^{(2)} = \begin{bmatrix} \sqrt{2a_1\mu_\nu} \\ \sqrt{2a_1\mu_\nu}[\sqrt{(2\pi)^2 + \mu_\nu^2 - 2\mu_\nu^2}] \end{bmatrix}$$

$$C^{(2)} = [1 \quad 0]$$

In general, the $(K+1)$ term of $r_a(\tau)$ is $a_k \exp(-\mu_\nu|\tau|) \cos 2\pi k\tau$ which has the bilateral transform $R_{k+1}(s)$, given by:

$$R_{k+1}(s) = \frac{2a_k\mu_\nu[-s^2 + (2k\pi)^2 + \mu_\nu^2]}{[(s+\mu_\nu)^2 + (2k\pi)^2][(-s+\mu_\nu)^2 + (2k\pi)^2]} \quad (5.39)$$

As before, the function $R_{k+1}(s)$ can be factored into two functions, $H_{k+1}(s)$ and $H_{k+1}(-s)$:

$$R_{k+1}(s) = \frac{\sqrt{2a_k\mu_\nu}[s + \sqrt{(2k\pi)^2 + \mu_\nu^2}]}{(s+\mu_\nu)^2 + (2\pi k)^2} \times \frac{\sqrt{2a_k\mu_\nu}[-s + \sqrt{(2k\pi)^2 + \mu_\nu^2}]}{(-s+\mu_\nu)^2 + (2k\pi)^2}$$

where

$$H_{k+1}(s) = \frac{\sqrt{2a_k\mu_\nu}[s + \sqrt{(2k\pi)^2 + \mu_\nu^2}]}{(s+\mu_\nu)^2 + (2k\pi)^2}$$

and the corresponding dynamical model is:

$$\dot{x}^{(k+1)} = A^{(k+1)}x^{(k+1)}(t) + B^{(k+1)}u^{(k+1)}(t)$$

$$y^{(k+1)}(t) = C^{(k+1)}x^{(k+1)}(t) \quad (5.40)$$

where

$$A^{(k+1)} = \begin{bmatrix} 0 & 1 \\ -(2k\pi)^2 + \mu_\nu^2 & -2\mu_\nu \end{bmatrix} \quad (5.41)$$

$$B^{(k+1)} = \begin{bmatrix} \sqrt{2a_k \mu_\nu} \\ \sqrt{2a_k \mu_\nu}[\sqrt{(2k\pi)^2 + \mu_\nu^2} - 2\mu_\nu^2] \end{bmatrix} \quad (5.42)$$

$$C^{(k+1)} = \begin{bmatrix} 1 & 0 \end{bmatrix} \quad (5.43)$$

It can be seen that the first term of $r_a(\tau)$ is modeled by Eq. (5.38), which is a first-order system, and the subsequent terms by (5.39), which is the second-order system. Thus, to model the $(J + 1)$ terms of $r_a(\tau)$, we need a $(2J + 1)$-order system. For example, suppose the function $r_a(\tau)$ has $(J + 1)$ terms; then we can incorporate the first- and second-order systems into a new system, whose parameters A, B, and C are obtained as follows:

$$A = \begin{bmatrix} -\mu_\nu & 0 & 0 & \cdots & 0 & 0 \\ 0 & 0 & 1 & \cdots & \cdot & \cdot \\ 0 & -[(2\pi)^2 + \mu_\nu^2] & -2\mu_\nu & \cdots & \cdot & \cdot \\ 0 & 0 & 0 & \cdots & \cdot & \cdot \\ \cdot & & & \cdots & & \cdot \\ \cdot & & & \cdots & & \cdot \\ \cdot & & & \cdots & & \cdot \\ 0 & 0 & 0 & \cdots & 0 & 1 \\ 0 & 0 & 0 & \cdots & -[(2\pi(J+1))^2 + \mu_\nu^2] & -2\mu_\nu \end{bmatrix} \quad (5.44)$$

$$B = \begin{bmatrix} \sqrt{2a_0 \mu_\nu} & 0 & \cdots & 0 \\ 0 & \sqrt{2a_1 \mu_\nu} & \cdot & 0 \\ 0 & \sqrt{2a_1 \mu_\nu}[\sqrt{(2\pi)^2 + \mu_\nu^2} - 2\mu_\nu^2] & \cdot & 0 \\ \cdot & \cdot & \cdot & 0 \\ \cdot & \cdot & \cdot & \sqrt{2a_J \mu_\nu} \\ 0 & 0 & & \sqrt{2a_J \mu_\nu}[\sqrt{(2\pi J)^2 + \mu_\nu^2} - 2\mu_\nu^2] \end{bmatrix} \quad (5.45)$$

$$C = \begin{bmatrix} 1 & 1 & 0 & 1 & 0 & \cdots & 1 & 0 \end{bmatrix} \quad (5.46)$$

Example 3

If in Example 2 only three terms of $r_a(\tau)$ are retained, i.e., $J = 2$, the resultant $r_a(\tau)$ can be written as:

$$r_a(\tau) = 6.1 \exp{-0.136(\tau)} \sum_{j=0}^{2} a_j \cos 2\pi\tau$$

If we use the Fourier series for $p(\tau)$, then a_0, a_1, and a_2 will be given as:

$$a_0 = 0.333; \quad a_1 = 0.405; \quad a_2 = 0.101$$

A plot of $r_a(\tau)$ is shown in Figure 5-2. The correlation term

$$6.1 a_0 \exp(-0.136|\tau|)$$

is modeled by:

$$\dot{x}_1 = -0.136 \, x_1(t) + 0.732 \, u_1$$

The second term in the correlation is modeled by:

$$\dot{x}_2 = x_3 + 0.82 \, u_2$$

$$\dot{x}_3 = -39.4 \, x_2 - 0.27 \, x_3 + 4.92 \, u_2$$

and the third term is modeled in a similar manner. The terms u_1, u_2, and u_3 represent independent white-noise terms, each with zero mean and correlation function $\delta(\tau)$, where δ is the Dirac delta function. The final results are:

$$A = \begin{bmatrix} -0.136 & 0 & 0 & 0 & 0 \\ 0 & 0 & 1 & 0 & 0 \\ 0 & -39.4 & -0.27 & 0 & 0 \\ 0 & 0 & 0 & 0 & 1 \\ 0 & 0 & 0 & -157.7 & -0.27 \end{bmatrix}$$

$$B = \begin{bmatrix} 0.743 & 0 & 0 \\ 0 & 0.820 & 0 \\ 0 & 4.92 & 0 \\ 0 & 0 & 0.410 \\ 0 & 0 & 5.04 \end{bmatrix}$$

$$C = [1 \quad 1 \quad 0 \quad 1 \quad 0]$$

Often, two-dimensional stationary correlation functions can be approximated by a combination of two-dimensional stationary correlation functions of the form:

$$R(x, i) = R(0,0) \exp\left(-\mu_h |x| - \mu_v |i|\right) \tag{5.47}$$

Because of the importance of $R(x, i)$ as given by Eq. (5.47), we shall discuss this special autocorrelation function below.

Calculating $r(\tau)$ (given by Eq. 5.22), one obtains:

$$r(\tau) = \frac{T-\gamma}{T} \exp(-\mu_h|\gamma| - \mu_v|i|)$$

$$+ \frac{\gamma}{T} \exp(-\mu_h|T-\gamma| - \mu_v|i+1|) \qquad (5.48)$$

where

$$\tau = iT + \gamma, \ 0 \leqslant \gamma \leqslant T$$

Now, let us define a risk function $\mathscr{R}(\cdot)$ such that

$$\mathscr{R}(r) = \int_0^{NT} [r(\tau) - r_a(\tau)]^2 \, d\tau \qquad (5.49)$$

and

$$r_a(\tau) = \sum_{j=0}^{J} a_j \exp(-\mu_v|\tau|) \cos \frac{2\pi j}{T} \tau \qquad (5.50)$$

We can select the coefficients a_j such that the risk function $\mathscr{R}(r)$ is minimized. For simplicity, we shall assume that $T = 1$. It can be shown that $\mathscr{R}(r)$ can be expressed by [4]:

$$\mathscr{R}(r) = \frac{1 - \exp(-2\mu_v N)}{1 - \exp(-2\mu_v)} \int_0^1 [r(\tau) - r_a(\tau)]^2 \, d\tau \qquad (5.51)$$

To minimize $\mathscr{R}(r)$, we must minimize:

$$\int_0^1 [r(\tau) - r_a(\tau)]^2 \, d\tau$$

Thus, the minimization of $\mathcal{R}(r)$ becomes a simple problem, and the risk function can be obtained from [4]. The procedure is to set the derivatives of $\mathcal{R}(r)$ with respect to a_j equal to zero, and the result can be obtained as follows:

$$a = \alpha^{-1} d \tag{5.52}$$

where α is a matrix, whose elements are given by:

$$\alpha_{kl} = \int_0^1 \exp(-2\mu_\nu |\tau|) \cos 2\pi k\tau \cos 2\pi l\tau \, d\tau \tag{5.53}$$

and d is a column vector, whose elements are given by:

$$d_k = \int_0^1 r(\tau) \exp(-\mu_\nu |\tau|) \cos 2\pi k\tau \, d\tau \tag{5.54}$$

Furthermore, the following properties can easily be established:

$$\int_0^1 [r(\tau) - r_a(\tau)]^2 \, d\tau = \int_0^1 r^2(\tau) - \int_0^1 r_a^2(\tau) \, d\tau \tag{5.55}$$

$$\int_0^1 r^2(\tau) \, d\tau = \lim_{j \to \infty} \int_0^1 r_a^2(\tau) \, d\tau \tag{5.56}$$

5.3.4 Design of a One-Step Predictor

Since we intend to utilize a digital computer for the estimation process, the model given by Eq. (5.26) is discretized, yielding:

$$x(k+1) = \bar{A}x(k) + \bar{B}u(k)$$

$$y(k) = \bar{C}x(k) + v(k) \tag{5.57}$$

In addition, the model given by Eq. (5.57) contains the observation noise element $v(k)$, which is assumed to be white, with mean zero and variance σ^2. The parameters $\bar{A}, \bar{B}$, and $\bar{C}$ are related to A, B, and C by:

$$\bar{A} = \exp\left(A\frac{T}{N}\right)$$

$$\bar{B}\bar{K}\bar{B}' = \int_0^{T/N} \exp\left(A\frac{T}{N}\right) \exp(-As) BKB'$$

$$\times \exp(-A's) \exp\left(A'\frac{T}{N}\right) ds$$

$$\bar{C} = C \tag{5.58}$$

where K and $\bar{K}$ are covariances of $u(t)$ and $u(k)$, respectively. The sampling interval utilized in the above discretization is chosen to be T/N. Thus, there will be N observations for each horizontal scan. Since there are N horizontal scan lines, the final discrete observation is on an $N \times N$ grid.

Example 4

Continuing Example 3, we obtain:

$$\bar{A} = \begin{bmatrix} 0.996 & 0 & 0 & 0 & 0 \\ 0 & 0.983 & 0.031 & 0 & 0 \\ 0 & -1.22 & 0.97 & 0 & 0 \\ 0 & 0 & 0 & 0.926 & 0.03 \\ 0 & 0 & 0 & -4.77 & 0.913 \end{bmatrix}$$

$$\bar{B}\bar{K}\bar{B}' = \begin{bmatrix} 0.02 & 0 & 0 & 0 \\ 0 & 0.02 & 0.12 & 0 \\ 0 & 0.12 & 0.60 & 0 \\ 0 & 0 & 0.01 & 0.07 \\ 0 & 0 & 0.07 & 0.49 \end{bmatrix}$$

$$\bar{C} = C = [1 \quad 1 \quad 0 \quad 1 \quad 0]$$

Utilizing the model given by Eq. (5.57) with parameters given by Eq. (5.58), a (one-step predictor) recursive estimator may be designed (see Chapter 4). The equations are given for the sake of completeness.

$$\hat{x}(k+1) = [\bar{A} - F(k)\bar{C}]\,\hat{x}(k) + F(k)\,y(k)$$

$$P(k+1) = [\bar{A} - F(k)\bar{C}]\,P(k)\,[\bar{A} - F(k)\bar{C}]' + \bar{B}\bar{K}\bar{B}' + F(k)F'(k)\,\sigma^2$$

$$F(k) = \bar{A}P(k)\bar{C}'\,[\bar{C}P(k)\bar{C}' + \sigma^2]^{-1} \tag{5.59}$$

The (one-step predicted) estimate of the image is, therefore,

$$\bar{C}\hat{x}(k) \doteq \hat{y}(k)$$

that is, $\hat{y}(k)$ is the best estimate of $y(k)$, obtained recursively in real time, where $y(k)$ is the observation associated with the grid point immediately ahead of the scanner position.

Example 5

The signal $y(k)$ is generated by using the image described in the preceding example and adding white noise with variance σ^2. Let us define a measure of signal-to-noise ratio by:

$$\rho \doteq \frac{\text{peak-to-peak variation of signal}}{\sigma}$$

The peak-to-peak variation of the image is 7.1. Two values of ρ are considered here, namely, 7.1/3 and 7.1/10; the corresponding values of $y(k)$ and their one-step predicted values $y(k)$ are shown in Figures 5-3a and 5-3b and 5-3a and 5-4b, respectively.

5.3.5 Implementation of Required Interpolation

It is clear that image enhancement, from the point of view of scanner output, represents an interpolation problem; i.e., it is desired to determine the best estimate of $y(k), 0 \leq k \leq N$, given the observation $y(0), y(1), \cdots, y(N)$. In general, the interpolation problem is far more complicated [13] than standard Kalman filtering. However, since for the image enhancement considered here the length of the data is fixed (N) and, furthermore, the observation is usually available for additional repeated processing, it is possible to obtain two one-step predicted values of $y(k)$, denoted by $\hat{y}(k)$ and $\tilde{y}(k)$, one by running the scanner in one direction starting, for example, at the top left corner of the picture and the other by running the scanner in the reverse direction starting at bottom right corner. Associated with these estimates are estimation error variances denoted by $\hat{\sigma}^2(k) = \overline{C\hat{P}(k)\,C'}$ and $\tilde{\sigma}^2(k) = \overline{C\tilde{P}(k)\,C'}$, respectively. The two estimates must be combined to yield the optimal interpolated (smoothed) value $y^*(k)$. Thus, a brief discussion of combining two estimators is warranted.

Suppose we are given two state estimates, $\hat{x}(t)$ and $\tilde{x}(t)$, of the same state variable $x(t)$. There are two cases to consider: either $\hat{x}(t)$ and $\tilde{x}(t)$ are correlated or they are uncorrelated. We shall combine only the case in which both are uncorrelated; i.e.,

$$E[x - \hat{x}][x - \tilde{x}]' = 0 \tag{5.60}$$

In this case the optimal estimate of x, denoted by $x^*(t)$, is given by:

$$x^* = P^*(\hat{P}^{-1}\hat{x} + \tilde{P}^{-1}\tilde{x}) \tag{5.61}$$

$$P^* = (\hat{P}^{-1} + \tilde{P}^{-1})^{-1} \tag{5.62}$$

where $\hat{P}$ and $\tilde{P}$ are the error covariances of $\hat{x}$ and $\tilde{x}$, respectively. Thus, applying Eqs. (5.60), (5.61), and (5.62) to obtain $\hat{y}(k) = C\hat{x}$ and $\tilde{y} = C\tilde{x}$ yields:

$$y^*(k) = \frac{\hat{\sigma}^2(k)}{\hat{\sigma}^2(k) + \tilde{\sigma}^2(k)}\hat{y}(k) + \frac{\tilde{\sigma}^2(k)}{\hat{\sigma}^2(k) + \tilde{\sigma}^2(k)}\tilde{y}(k) \tag{5.63}$$

Example 6

Considering the preceding example, the covariance $P(k)$ in Eq. (5.59) nearly reaches its steady-state value in about two or three scan lines. Consequently, $\hat{\sigma}(k) \approx \tilde{\sigma}(k)$ for most of the picture, and Eq. (5.63) reduces to:

$$y^*(k) \approx \frac{1}{2} [\hat{y}(k) + \tilde{y}(k)] \tag{5.64}$$

Equation (5.64) was implemented, and the results for $\rho = 7.1/3$ and $7.1/10$ appear in Figures 5-3c and 5-4c, respectively.

Careful observation of Figures 5-3b and 5-3c (or 5-4b and 5-4c) reveals a consistent vertical correlation, which is attributed to the approximation of $r(\tau)$ by transposing the original picture and re-evaluating $y^*(k)$. The two values are

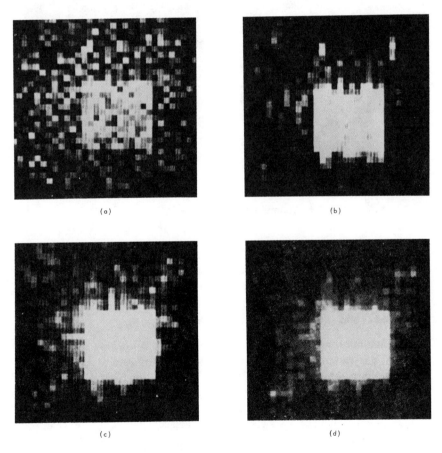

Fig. 5-3. Observation and Estimates for $\rho = 7/3$

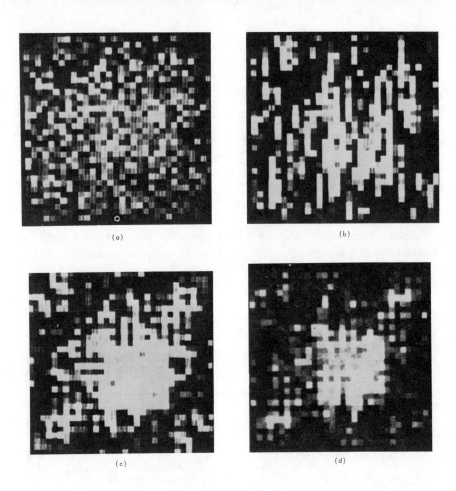

Fig. 5-4. Observation and Estimates for $\rho = 7/10$

averaged and are represented in Figures 5-3d and 5-4d for corresponding values of ρ. In what follows the approximation is further improved.

5.4 PARTIAL RANDOMIZATION

The randomization of ξ over the period T has the intuitive appeal that all points of the picture are weighted equally. While the results concerning this particular approximation to a certain subclass of nonstationary correlation functions have indeed been gratifying, it may lead to some shortcomings. For example, the extreme right edge of a scanned line and the extreme left edge of the next line would be weighted as two adjacent points of a line. In order to enhance the quality of our approximation, we shall discuss the idea of partial randomization, which assumes σ is randomly distributed over subintervals of $[0,T]$. Intuitively, it can be seen that the more the number of subdivisions, the closer we approximate the correlation function of the scanner output. Thus, we shall subdivide the image in the manner given below.

Let us subdivide $[0, T]$ into M parts such that:

$$0 = T_0 < T_1 < T_2 < \ldots < T_M = T \tag{5.65}$$

Let Δ_η be defined as:

$$\Delta_\eta \doteq T_\eta - T_{\eta-1}, \text{ for } \eta = 1,2,\ldots,M \tag{5.66}$$

Now for given $t = jT + \sigma$, where $\sigma \in [T_{\eta-1}, T]$, in a manner to that before, let $q_\eta(t)$ be a random variable such that:

$$q_\eta(t) \doteq s(jT + \xi) \tag{5.67}$$

where ξ is assumed to be uniformly distributed over $[T_{\eta-1}, T_\eta]$ for $\eta = 1,2,\ldots,M$ and $q_\eta(t)$ is not defined elsewhere. Now we shall prove the following theorem.

Theorem 3

The random process $q_\eta(t)$ defined by Eq. (5.67) is stationary.

Proof

It is easy to verify that:

$$Eq_\eta(t) = 0$$

by construction of $q_\eta(t)$.

Next, we shall prove that $Eq_\eta(t)q_\eta(t + \tau)$ is a function of τ (or, equivalently, γ) only. $Eq_\eta(t) q_\eta(t + \tau)$ can be calculated as follows:

$$Eq_\eta(t) q_\eta(t + \tau) = E_\xi E_s s(jT + \xi) s(iT + jT + \xi + \gamma) \tag{5.68}$$

where (see 5.17):

$$\tau = iT + \gamma, \quad i = 0,\pm1,\pm2,\ldots, \quad 0 \leq t + \tau \leq NT \tag{5.69}$$

and ξ is uniformly distributed over $[T_{\eta-1}, T_\eta]$ and $|\gamma| \leq \Delta_\eta$. For $1 \leq \eta < M$, it is clear that $\xi + \gamma < T$. Utilizing (5.19), we obtain:

$$Eq_\eta(t) q_\eta(t + \tau) = \frac{1}{\Delta_k} \int_{T_{\eta-1}}^{T_\eta} E_s s(jT + \sigma) s(iT + jT + \sigma + \gamma) d\sigma$$

$$= \frac{1}{\Delta_\eta} \int_{T_{\eta-1}}^{T_\eta} R(\gamma, i) d\sigma = \frac{T_\eta - T_{\eta-1}}{\Delta_\eta} R(\gamma, i) = R(\gamma, i)$$

(5.70)

However, for $\xi \in [T_{M-1}, T_M] = [T_{M-1}, T]$, $\xi + \gamma$ may no longer be less than T. Utilizing Eq. (5.19) once more, we get:

$$Eq_M(t) q_M(t + \tau) = \frac{1}{\Delta_M} \int_{T_{M-1}}^{T} E_s s(jT + \sigma) s(jT + iT + \sigma + \gamma) d\sigma$$

$$= \frac{1}{\Delta_M} \int_{T_{M-1}}^{T-\gamma} R(\gamma, i) d\sigma + \frac{1}{\Delta_M} \int_{T-\gamma}^{T} R(\gamma - T, i + 1) d\sigma$$

$$= \frac{\Delta_M - \gamma}{\Delta_M} R(\gamma, i) + \frac{\gamma}{\Delta_M} R(\gamma - T, i + 1) \quad (5.71)$$

where $|\gamma| < \Delta_M$, which concludes the proof.

Let S_η be defined as follows:

$$S_\eta \doteq \{t : t \in (jT, jT + \Delta_\eta), j = 0, 1, \ldots, N - 1\} \quad (5.72)$$

where Δ_η is defined by Eq. (5.56). Hence, the entire picture consists of the collection of partitions $S_1, S_2, \ldots, S_M$, as shown in Figure 5-5.

Let $\theta(t)$ be the observation given by:

$$\theta(t) = s(t) + v(t), \ t \in S_1 \quad (5.73)$$

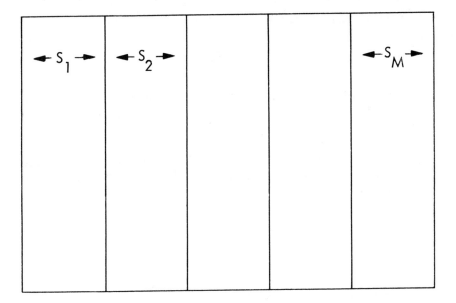

Fig. 5-5. Partitioned Image

where $v(t)$ is the white noise of zero mean and variance σ^2. Now we can state a very important result via a theorem.

Theorem 4

The second-order statistical information of $s(t)$ and $\theta(t)$ for $t \in S_1$ is sufficient for obtaining the best linear mean square estimate of $s(t)$ denoted as $\hat{s}(t)$, given the observation $\theta(t)$, $t \in S_1$. The optimal solution is unique and independent of the particular generating model of signal process $s(t)$.

Proof

Let $L(\alpha(\tau), t)$ be the operator defined by:

$$L(\alpha(\tau), t)\,\theta(\tau) \doteq \int_0^{T_1} \alpha(\tau)\theta(\tau)\, d\tau$$

$$+ \int_T^{T+T_1} \alpha(\tau)\theta(\tau)\, d\tau + \ldots + \int_{jT}^{t} \alpha(\tau)\theta(\tau)\, d\tau \tag{5.74}$$

where $\alpha(\tau)$ is a scalar function. We are interested in minimizing:

$$E[s(t) - \hat{s}(t)]^2, \ t \in S_1 \tag{5.75}$$

where $\hat{s}(t)$ is restricted to a linear function of the observation $\theta(\tau)$, $\tau \leq t$ with both τ and t belonging to S_1. Consequently, $\hat{s}(t)$ has the form given by Eq. (5.74).

It is desired to find that $\alpha(\tau)$, denoted by $\alpha^0(\tau)$, which will minimize (5.75). Using the ideas of calculus of variations [13], let $\alpha^{00}(\tau)$ be any arbitrary function of τ and ϵ be an arbitrary small scalar. Letting

$$\alpha(\tau) = \alpha^0(\tau) + \epsilon \alpha^{00}(\tau)$$

and substituting this in Eq. (5.75) yields:

$$E[s(t) - L(\alpha^0(\tau) + \epsilon \alpha^{00}(\tau), t) \theta(\tau)]^2 \tag{5.76}$$

where the expectation is over t and τ.

If $\alpha^0(\tau)$ yields the minimum value for Eq. (5.74), then the coefficients of the term in ϵ in the expansion of Eq. (5.76) must be zero, since ϵ can be chosen small and with arbitrary sign. It follows that:

$$EL(\alpha^{00}(\tau), t)\left[\theta(\tau)(s(t) - \hat{s}(t))\right] = 0$$

Or, in the expanded form,

$$E\left[\int_0^T \alpha^{00}(\tau)\theta(\tau)[s(t) - \hat{s}(t)]\, d\tau + \ldots + \int_{jT}^t \alpha^{00}(\tau)\theta(\tau)[s(t) - \hat{s}(t)]\, d\tau\right] = 0$$

Since the above equation must be satisfied for any $\alpha^{00}(\tau)$, we must necessarily have:

$$E[s(t) - \hat{s}(t)]\, \theta(\tau) = 0, \text{ for } 0 \leqslant \tau \leqslant T_1$$

$$E[s(t) - \hat{s}(t)]\, \theta(\tau) = 0, \text{ for } T \leqslant \tau \leqslant T + T_1$$

$$\vdots$$

$$E[s(t) - \hat{s}(t)]\, \theta(\tau) = 0, \text{ for } jT \leqslant \tau t$$

which is the orthogonality principle, that is,

$$E[s(t) - \hat{s}(t)]\, \theta(\tau) = 0, \text{ for } \tau, t \in S_1 \text{ and } \tau \leqslant t \tag{5.77}$$

The solution of Eq. (5.20) yields the optimal solution $\hat{s}(t)$. Equivalently, Eq. (5.77) may be written as:

$$Es(t)\, \theta(\tau) = E\hat{s}(t)\, \theta(\tau)$$

where $\hat{s}(t)$ is given by (5.74). Hence, we have:

$$Es(t)\, \theta(\tau) = \int_0^{T_1} \alpha^0(\tau)\, E\theta(\tau)\, \theta(t)\, dt + \int_T^{T+T_1} \alpha^0(\tau)\, E\theta(\tau)\, \theta(t)\, dt + \ldots$$

$$+ \int_{jT}^{t} \alpha^0(\tau) E\theta(\tau)\theta(t)\, dt$$

which implies that the optimal solution depends on the second moment statistics of $s(t)$ and $\theta(t)$ over S_1 only.

Example 7

Consider a square picture subdivided into a 32 × 32 grid. Let $T = 1$ second and $\nu = 1$. The signal is a 20 × 20 square starting from the thirteenth row and the first column. Let m and n represent specific rows and columns, respectively. The above signal is represented by the brightness level $b(m, n) = 1.56$, where a signal exists, and -1 otherwise, resulting in a zero mean sample function. As a first approximation, let us choose:

$$R(z, i) = \alpha \exp(-\mu_h |z| - \mu_v |i|)$$

where α, μ_h, and μ_v are to be determined. Computation of the sample power gives rise to $\alpha = R(0,0) \approx 1.56$. The correlation between two adjacent grid points is calculated as 1.394, which is the value for $R(1/32,0)$ or $R(0,1)$. Hence,

$$R(z, i) = 1.56 \exp(-3.44|z| - 0.107|i|)$$

Example 8

Let us partition the above picture into three parts S_1, S_2, and S_3, where S_n is given by Eq. (5.72). We subdivide [0,1] as follows:

$$0 = T_0 < T_1 < T_2 < T_3 = 1$$

with

$$\Delta_1 = \Delta_3 = \frac{11}{32} \quad \text{and} \quad \Delta_2 = \frac{10}{32}$$

Then, $Eq_1(t) \, q_1(t + \tau)$ for t and $t + \tau \in S_1$ can be calculated by utilizing Eq. (5.70) and is given by:

$$Eq_1(t) \, q_1(t + \tau) = 1.56 \exp(-3.44|\gamma| - 0.107|i|)$$

Similarly, $Eq_2(t) \, q_2(t + \tau)$ for $t, t + \tau \in S_2$ is given by:

$$Eq_2(t) \, q_2(t + \tau) = 1.56 \exp(-3.44|\gamma| - 0.107|i|)$$

$Eq_3(t)\,q_3(t+\tau)$ for $t, t+\tau \in S_3$ can be calculated from (5.71) and is given by:

$$Eq_3(t)q_3(t+\tau) = 1.56\left[\frac{\Delta_3 - \gamma}{\Delta_3}\exp(-3.44|\gamma| - 0.107|i|)\right.$$

$$\left. + \frac{\gamma}{\Delta_3}\exp(-3.44|\gamma - 1| - 0.107|i+1|)\right]$$

5.4.1 Dynamic Modeling of Image Statistics

Now, for any $1 \leq \eta \leq M$, we wish to derive a differential equation model whose solution has an autocorrelation function approximating $Eq_\eta(t)\,q_\eta(t+\tau)$. We subsequently intend to utilize a Kalman estimator for each η, whenever the signal $q_\eta(t)$ is contaminated by additive white noise. But, from Theorem 3, the linear minimum mean square estimate $\hat{q}_\eta(t)$ is independent of the particular dynamic model generating the signal process $q_\eta(t)$. Hence, it is sufficient to devise any stationary correlation function which matches the first two moments of $q_\eta(t)$ for $t \in S_\eta$.

Again, without any loss of generality, we let $\eta = 1$, since the analysis would be similar for $\eta > 1$. Let the dynamic model

$$\dot{x} = \overline{A}_1 x(t) + \overline{B}_1 u(t)$$

$$y(t) = \overline{C}_1 x(t)$$

(5.78)

be such that its output correlation function denoted as $\phi_1(\tau)$ satisfies:

$$\phi_1(\tau) = Eq_1(t)\,q_1(t+\tau), \text{ for } t, t+\tau \in S_1 \quad (5.79)$$

where $x_1(t)$ is an n-dimensional vector, $u(t)$ is a white noise vector, and $y(t)$ is the scalar signal whose autocorrelation function is $\phi_1(\tau)$. The procedure followed is to present an approximation to $\phi_1(\tau)$, denoted as $\phi_{1a}(\tau)$, as a sum of terms such that each term can easily be modeled. The procedure has been discussed; however, we shall repeat it for the sake of completeness.

Let us decompose $\phi_1(\tau)$ into the product of two functions $\xi_1(\tau)$ and $\phi_1(\tau)/\xi_1(\tau)$, where $\xi_1(\tau)$ is chosen to satisfy $\xi_1(iT) = R(0, i)$ for all i, and $\xi_1(\tau)$ is taken to be a combination of non-negative exponentials, i.e.,

$$\xi_1(\tau) = \sum_{i=1}^{I} l_i \exp(-\lambda_i |\tau|)$$

The function $p_1(\tau)$ is chosen to be a periodic function approximating $\phi_1(\tau)/\xi_1(\tau)$. The approximate correlation function is then,

$$\phi_{1a}(\tau) = \xi_1(\tau) p_1(\tau) \tag{5.80}$$

A natural candidate for $p_1(\tau)$ is to choose $p_1(\tau)$ as:

$$p_1(\tau) = \sum_{j=0}^{J} a_j \cos \frac{2\pi j}{T} \tau \tag{5.81}$$

Hence, an element of the correlation function $p_1(\tau)$ has the form:

$$l_i a_j \exp(-\lambda_i |i|) \cos \frac{2\pi j}{T} \tau$$

and there are $(J + 1) I$ such elements. A differential equation model with white noise input can simply be constructed to model each of these terms. Each will be a second-order system except those corresponding to $j = 0$, which will be of the first order. If the white noise terms are assumed to be mutually independent, the collection of all these differential equations defines $\overline{A}_1, \overline{B}_1$, and $\overline{C}_1$ and represents the desired model for $\phi_{1a}(\tau)$.

Example 9

In Example 8, due to the exponential nature of $R(z, i)$, we choose:

$$\xi_1(\tau) = R(0,0) \exp(-0.107|\tau|)$$

Only three terms in (5.81) are retained; that is, $J = 2$. The resultant $\phi_{1a}(\tau)$ is:

$$\phi_{1a}(\tau) = 1.56 \exp(-0.107|\tau|) \sum_{j=0}^{2} a_j \cos 2\pi j \tau$$

where $a_0 = 0.396$, $a_1 = 0.445$, and $a_2 = 0.0131$. The autocorrelation term:

$$1.56 a_0 \exp(-0.107|\tau|)$$

is modeled by $x_1(t)$, where

$$\dot{x}_1 = 0.107 x_1 + 0.365 u_1$$

The second term in the correlation $\phi_{1a}(\tau)$ is modeled by:

$$\dot{x}_2 = x_3 + 0.368 u_2$$

$$\dot{x}_3 = -39.4 x_2 - 0.214 x_3 + 2.42 u_2$$

The third term is modeled in a similar manner. The u_1, u_2, and u_3 represent independent white noise terms, each with zero mean and correlation function $\delta(\tau)$, where δ is the Dirac delta function. The final results are:

$$\overline{A}_1 = \begin{bmatrix} -0.107 & 0 & 0 & 0 & 0 \\ 0 & 0 & 1 & 0 & 0 \\ 0 & -39.4 & -0.214 & 0 & 0 \\ 0 & 0 & 0 & 0 & 1 \\ 0 & 0 & 0 & -157.7 & -0.214 \end{bmatrix}$$

$$\overline{B}_1 = \begin{bmatrix} 0.365 & 0 & 0 \\ 0 & 0.386 & 0 \\ 0 & 2.42 & 0 \\ 0 & 0 & 0.065 \\ 0 & 0 & 0.834 \end{bmatrix}$$

$$\overline{C}_1 = [1 \quad 1 \quad 0 \quad 1 \quad 0]$$

The dynamic model generating the signal process $s_2(t)$ is identical to that of $s_1(t)$. However, the dynamic model corresponding to the signal process $s_3(t)$ is given by:

$$\dot{\overline{x}} = \overline{A}_3 \overline{x}(t) + \overline{B}_3 u(t)$$

$$\overline{y}(t) = \overline{C}_3 \overline{x}(t)$$

where

$$\overline{A}_3 = \overline{A}_1$$

and

$$\overline{B}_3 = \begin{bmatrix} 0.334 & 0 & 0 \\ 0 & 0.334 & 0 \\ 0 & 2.1 & 0 \\ 0 & 0 & 0.11 \\ 0 & 0 & 2.65 \end{bmatrix}$$

$$\overline{C}_3 = \overline{C}_1$$

5.4.2 Design of Estimator

From Eqs. (5.70) and (5.71), it follows that two different dynamic models corresponding to the correlation functions exist, one for $1 \leq k < M$ and the other for $k = M$. In what follows we intend to utilize a digital computer for the estimation process. The model corresponding to $1 \leq k < M$ is given by Eq. (5.78). For $k = M$, let the corresponding dynamic model be given by:

$$\dot{x} = \overline{A}_M \overline{x}(t) + \overline{B}_M u(t)$$

$$\overline{y}(t) = \overline{C}_M \overline{x}(t)$$
(5.82)

i.e., the dynamic model generates the signal process $\overline{y}(t)$. Let us assume that both dynamic models, given by Eq. (5.78) and Eq. (5.82), are of the same dimensions. Discretizing Eq. (5.87) yields:

$$x(k+1) = A_1 x(k) + B_1 u(k)$$

$$y(k) = C_1 x(k) + v(k)$$
(5.83)

In addition, the model given by Eq. (5.83) contains the observation (background) noise element $v(k)$, which is assumed to be white of zero mean and variance σ^2. The parameters A_1, B_1, C_1 are related to $\overline{A}_1, \overline{B}_1$, and $\overline{C}_1$ by:

$$A = \exp\left(\overline{A}_1 \frac{T}{N}\right)$$

$$B_1 K_1 B_1' = \int_0^{T/N} \exp\left(\overline{A}_1 \frac{T}{N}\right) \exp(-\overline{A}_1 s) \overline{B}_1 \overline{K}_1 \overline{B}_1' \exp(-\overline{A}_1' s) \exp\left(\overline{A}_1' \frac{T}{N}\right) ds$$

$$C_1 = \overline{C}_1$$
(5.84)

where exp is the exponent, and K_1 and $\overline{K}_1$ are covariances of $u(k)$ and $u(t)$, respectively. We discretize Eq. (5.82) in the same manner. Let

$$x(k+1) = A_M x(k) + B_M u(k)$$

$$y(k) = C_M x(k) + v(k)$$
(5.85)

with its corresponding parameters given by:

$$A_M = \exp\left(\overline{A}_M \frac{T}{N}\right)$$

$$B_M K_M B'_M = \int_0^{T/N} \exp\left(\overline{A}_M \frac{T}{N}\right) \exp(-A_M s)\overline{B}_M \overline{K}_M \overline{B}'_M \cdot$$
$$\exp(-\overline{A}'_M s) \exp\left(\overline{A}'_M \frac{T}{N}\right) ds \qquad (5.86)$$

$$C_M = \overline{C}_M$$

Example 10

In this example, let $\hat{y}(k)$ denote the estimate of $C_1 x(k)$ or $C_M \bar{x}(k)$. Every k can be written as $k = 32i + j$, for $i = 1,2,\ldots,N$ and $1 \leq j < 32$, where i is the ith scanned line and j determines the position on the ith scanned line. Continuing Examples 8–10, we can see that the start of the three vertical strips corresponds to the values of $j = 1$, 11, or 21. For $1 \leq j \leq 21$, we utilize model Eq. (5.83), since the values of η would be either 1 or 2. For other values of j, we utilize model Eq. (5.85). Now for the values of $j = 1$, 11, and 21, the best linear mean square estimate of $y(k)$ must be the optimal combination of $\hat{y}(k)$ and $\hat{y}(k-32)$, where the two estimates use a portion of the observation twice. However, the overlapped portion of the observation is very small, and the optimality will not be significantly affected by assuming the estimators to be independent.

The formula for combining two independent estimates $\hat{x}$ and $\tilde{x}$ of the same state variable x to obtain a combined estimate x^* with its associated covariance error given by (see Chapter 4):

$$x^* = P^*(\hat{P}^{-1}\hat{x} + \tilde{P}^{-1}\tilde{x}) \qquad (5.87)$$

$$P^* = (\hat{P}^{-1} + \tilde{P}^{-1})^{-1} \qquad (5.88)$$

where $\hat{P}$ and $\tilde{P}$ are the error covariances of $\hat{x}$ and $\tilde{x}$, respectively, thus, applying Eqs. (5.87) and (5.88) to $\hat{y}(k) = C\hat{x}(k)$ and $\tilde{y}(k) = C\tilde{x}(k)$ yields:

$$y^*(k) = \frac{\tilde{\sigma}^2(k)}{\hat{\sigma}^2(k) + \tilde{\sigma}^2(k)} \hat{y}(k) + \frac{\hat{\sigma}^2(k)}{\hat{\sigma}^2(k) + \tilde{\sigma}^2(k)} \tilde{y}(k)$$

where y^* denotes the combined estimate for $y(k)$. Continuing Example 9, we obtain:

$$A_1 = A_M = \begin{bmatrix} 0.996 & 0 & 0 & 0 & 0 \\ 0 & 0.983 & 0.031 & 0 & 0 \\ 0 & -1.223 & 0.970 & 0 & 0 \\ 0 & 0 & 0 & 0.926 & 0.03 \\ 0 & 0 & 0 & -4.75 & 0.93 \end{bmatrix}$$

$$B_1 K_1 B'_1 = \begin{bmatrix} 0 & 0 & 0 & 0 & 0 \\ 0 & 0.01 & 0.03 & 0 & 0 \\ 0 & 0.03 & 0.15 & 0 & 0 \\ 0 & 0 & 0 & 0 & 0 \\ 0 & 0 & 0 & 0 & 0.01 \end{bmatrix}$$

$$B_M K_M B'_M = \begin{bmatrix} 0 & 0 & 0 & 0 & 0 \\ 0 & 0 & 0.02 & 0 & 0 \\ 0 & 0.02 & 0.11 & 0 & 0 \\ 0 & 0 & 0 & 0 & 0.02 \\ 0 & 0 & 0 & 0.02 & 0.14 \end{bmatrix}$$

$$C_1 = C_M = [1 \quad 1 \quad 0 \quad 1 \quad 0]$$

Utilizing the models given by Eqs. (5.83) and (5.85) with their corresponding parameters given by Eqs. (5.84) and (5.86) respectively, a (one-step predictor) recursive estimator for each system may now be designed. The equations for (5.83) are given for the sake of completeness:

$$\hat{x}(k+1) = [A_1 - F_1(k)C_1]\hat{x}(k) + F_1(k)y(k)$$

$$P_1(k+1) = [A_1 - F_1(k)C_1]P_1(k)[A_1 - F_1(k)C_1]'$$
$$+ B_1 K B_1' + F_1(k)F_1'(k)\sigma^2$$

$$F_1(k) = AP_1(k)C_1'\,[C_1 P_1(k)C_1' + \sigma^2]^{-1}$$

A similar set of equations exists for (5.85), the only difference being a change of subscripts from 1 to M.

The one-step predicted estimate $\hat{y}(k)$ of $y(k)$ is found recursively in real time. $y(k)$ is the observation associated with the grid immediately ahead of scanner position.

Example 11

The signal $\bar{y}(t)$ or $y(k)$ is generated by using the image described in the preceding example and by adding white noise with variance σ^2. The peak-to-peak variation is 2.56. Let us select as a measure of signal-to-noise ratio:

$$\rho \doteq \frac{\text{peak-to-peak variation of signal}}{\sigma}$$

A value of ρ of 2.56/10, which represents a very noisy image, was utilized. Figure 5-6 represents the uncontaminated image, where the corresponding values

Fig. 5-6. Uncontaminated Image

of $y(k)$ and their one-step predictors are shown in Figures 5-7a and 5-7b, respectively.

Example 12

Since the length of data is fixed and the observation is available for additional repeated processing, it is possible to obtain two one-step predicted values of $y(k)$, denoted as $\hat{y}(k)$ and $\tilde{y}(k)$, starting from the top left corner of the image and the other by running the scanner in the reverse direction starting at the bottom right corner. Associated with these estimates are estimation error variances denoted by $\hat{\sigma}^2(k) = C\hat{P}(k)C'$ and $\tilde{\sigma}^2(k) = C\tilde{P}(k)C'$, respectively. The result of combining the two estimates for $\rho = 2.56/10$ appears in Figure 5-7c.

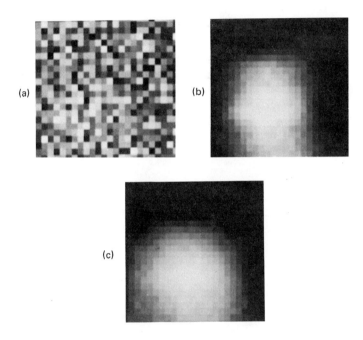

Fig. 5-7. Observation and Estimates for $\rho = 2.56/10$

5.5 CONCLUSIONS

The role of recursive (Kalman) filtering in image processing has been established. The procedure is applicable to those images characterized statistically by their mean and correlation function. A recursive estimation approach is very desirable due to its computational advantages. The effectiveness and the computational simplicity of our method to enhance contaminated images have been demonstrated via examples.

APPENDIX A
DIRAC DELTA FUNCTION

We have often seen the "delta" function $\delta(x)$ described as:

$$\int_{-\infty}^{\infty} \delta(x)\, dx = 1,\ \delta(x) = 0,\text{ for } x \neq 0$$

We must point out that $\delta(x)$ is not a function, but a mathematical symbol. We shall discuss the definition of $\delta(x)$ below.

Definition 1

A function $\phi(t)$, which is differentiable infinitely many times, is said to belong to class C or, symbolically, $\phi \in C$ if the following condition is satisfied:

$$\lim_{|t| \mapsto \infty} [t^i\, \phi^{(j)}(t)] = 0,\text{ for all } i \text{ and } j \geq 0$$

Note that $\phi^{(j)}(t)$ denotes the jth derivative.

Now we need to define another expression.

Definition 2

The sequence of functions $g_1(t), g_2(t), \ldots$ of class C is said to be regular if for any function $\phi(t) \in C$:

$$\lim_{n \to \infty} (g_n, \phi) \triangleq \lim_{n \to \infty} \int_{-\infty}^{\infty} g_n(t) \phi(t) \, dt$$

is finite.

Example

Consider the sequence

$$\left\{ \sqrt{\frac{n}{\pi}} \exp(-nt^2) \right\} = \{g_n(t)\}$$

The function $g_n(t)$ is of class C and

$$\lim_{n \to \infty} \sqrt{\frac{n}{\pi}} \exp(-nt^2) \to \infty$$

However, for any function $\phi \in C$,

$$\lim_{n \to \infty} (g_n, \phi)$$

is finite.

Definition 3

Two regular sequence of functions $\{g_n(t)\}$ and $\{h_n(t)\}$ are equivalent if

$$\lim_{n \to \infty} (g_n(t), \phi) = \lim_{n \to \infty} (h_n(t), \phi)$$

We shall denote $g_n \sim h_n$.

For example,

$$\left\{\sqrt{\frac{n}{\pi}} \exp(-nt^2)\right\} \quad \text{and} \quad \left\{\frac{1}{\sqrt{2\pi n}} \exp\left(\frac{-t^2}{2n^2}\right)\right\}$$

are equivalent, even though the functions are not equal to each other.

Definition 4

If the limit of $\{g_n(t)\}$ (with respect to a function $\phi \in C$) converges to a function g, i.e.,

$$(g, \phi) = \lim_{n \to \infty} (g_n, \phi)$$

then g is called a generalized function and $g \sim \{g_n\}$. A generalized function denoted by u is called a unit step function if

$$(u, \phi) = \lim_{n \to \infty} \int_{-\infty}^{\infty} u_n(t) \phi(t) \, dt = \int_{-\infty}^{\infty} u(t) \phi(t) \, dt$$

for all classes of $\{u_n(t)\}$, where

$$u(t) \triangleq \begin{cases} 1, & \text{if } t > 0 \\ 0, & \text{if } t \leq 0 \end{cases}$$

Example

The sequence

$$u_n(t) = \begin{cases} \exp\left[-\frac{1}{n}\left(\frac{k}{t} + t^2\right)\right], & \text{if } t > 0 \\ 0, & \text{if } t \leq 0 \end{cases}$$

represents a generalized unit step function.

Definition 5

The unit impulse or Dirac delta function $\delta(t)$ is defined as:

$$\delta \sim \{u_n'(t)\}$$

That is,

$$(\delta, \phi) = \lim_{n \to \infty} (u_n', \phi)$$

It should be emphasized that $\delta(t)$ is merely a symbol representing the total class of equivalent regular sequences $\{u_n'(t)\}$. Hence,

$$\int_{-\infty}^{\infty} \delta(t) \phi(t) \, dt = \lim_{n \to \infty} \int_{-\infty}^{\infty} u_n'(t) \phi(t) \, dt$$

Example

The sequence $\{u_n'(t)\}$ given by:

$$u_n'(t) = \begin{cases} \left[\dfrac{k}{t^2} - 2t\right] \exp\left[-\dfrac{1}{n}\left(\dfrac{k}{t} + t^2\right)\right], & \text{if } t > 0 \\ 0, & \text{if } t < 0 \end{cases}$$

is only one sequence which represents $\delta(t)$. Other sequences are:

$$\left\{\sqrt{\dfrac{n}{\pi}} \exp(-nt^2)\right\}, \quad \left\{\dfrac{1}{\sqrt{2\pi n}} \exp\left(\dfrac{-t^2}{2n^2}\right)\right\}, \text{ etc.}$$

The following important properties of $\delta(t)$ will hold:

(1) $\displaystyle\int_{\alpha < 0}^{\beta > 0} \delta(t) f(t) \, dt = f(0)$

where f is differentiable over the interval $\alpha \leqslant t \leqslant \beta$.

(2) $\int_{\alpha<0}^{\beta>0} f(t)\,\delta(t-a)\,dt = f(a)$

Both equations can be proven from the definition and utilizing the integration by part.

(3) $\int_{\alpha<0}^{\beta>0} \delta(t)\,dt = 1,\ \delta(t) = 0,\ t \neq 0$

APPENDIX B

VECTOR SPACES AND MATRICES

Definition 1

Let V be a set; then V is called a linear vector space over the real or the complex field if the following rules are satisfied:

(1) If $x \in V, y \in V$, then $x + y \in V$
(2) $(x + y) + z = x + (y + z)$
(3) There exists a "zero" vector $0 \in V$ such that $x + 0 = 0 + x = x$ for every $x \in V$
(4) For every $x \in V$, there exists another $x^- \in V$ such that $x + x^- = 0$
(5) $x + y = y + x$ for all x and $y \in V$

There exists a set of scalars (either real R or complex C) denoted by Greek letters such that:

(6) $(\alpha + \beta) x = \alpha x + \beta x$ (Distributive Law)
(7) $\alpha(x + y) = \alpha x + \alpha y$ (Distributive Law)
(8) $(\alpha\beta)(x) = \alpha(\beta x)$ (Associative Law)
(9) $1 \cdot x = x$
(10) $0 \cdot x = 0$

It can be shown that a set V is a vector space iff for any $x, y \in V$ and any scalars α and β, $\alpha x + \beta y \in V$. The most important example of the vector space is R^n.

Definition 2

Let V and W be linear vector spaces over the same field of scalars, and let T be a mapping (transformation) $V \to W$ such that:

(1) $T(x + y) = Tx + Ty$ for all x and $y \in V$

(2) $T(\alpha x) = \alpha Tx$ for all $x \in V$ and all scalars α

Then T is said to be linear.

Definition 3

A set of vectors $\{x_1, x_2, \ldots, x_n\}$ is a basis in V if:

(1) The set is linearly independent (no x_i's can be written as a linear combination of the other vectors).

(2) They generate the vector space V, i.e., every $x \in V$ can be written as a linear combination of $x_1, x_2, \ldots, x_3$.

Definition 4

The number of linearly independent vectors n in Definition 3 is called the dimension of the vector space V.

(1) $A(x + y) = A(x) + T(y)$, for any x and $y \in V$

(2) $A(\alpha x) = \alpha A(x)$, for any scalar α and $x \in V$

Definition of Matrices

Let $\{e_1, e_2, \ldots, e_n\}$ be a basis in V and $\{f_1, f_2, \ldots, f_m\}$ be a basis in W, and assume A is a linear transformation

$$A: V \to W$$

Then $A(e_j) \in W$ for all $j = 1, 2, \ldots, n$ which implies that:

$$A(e_j) = \sum_{i=1}^{m} a_{ij} f_i \qquad (B.1)$$

or in the expanded form:

$$A(e_1) = a_{11} f_1 + a_{21} f_2 + \ldots + a_{m1} f_m$$

$$A(e_2) = a_{12} f_1 + a_{22} f_2 + \ldots + a_{m2} f_m$$

$$\vdots$$

$$A(e_n) = a_{1n} f_1 + a_{2n} f_2 + \ldots + a_{mn} f_m$$

Definition 5

Now the matrix of A denoted by M_A with respect to the above basis is defined as:

$$M_A = \begin{bmatrix} a_{11} & a_{12} & \cdots & a_{1n} \\ a_{21} & a_{22} & \cdots & a_{2n} \\ \vdots & & & \\ a_{m1} & a_{m2} & \cdots & a_{mn} \end{bmatrix}_{m \times n} = [a_{ij}]_{m \times n}$$

Thus the matrix $[a_{ij}]_{m \times n}$ depends on the linear transformation A as well as the bases in V and W.

Let A and B be linear transformations with respect to the same spaces; then the reader is advised to prove the following properties:

(1) $M_A + M_B = [a_{ij} + b_{ij}]_{m \times n}$

(2) $M_{\alpha A} = \alpha M_A$

where M_A and M_B are the matrices with respect to the operators A and B, α is a scalar, and $[a_{ij} + b_{ij}]_{m \times n}$ is the matrix with respect to the operator $A + B$.

Definition 6

(1) If $Ax = x$ for every x, then the operator is called the identity and is denoted by I.

(2) A is a zero operator if $Ax = 0$ for every $x \in V$.

(3) If $V = W$, then A is said to be invertible iff $Ax_1 = Ax_2$ implies $x_1 = x_2$ and, for every $y \in V$, there exists an $x \in V$ such that $Ax = y$. If A is not invertible it is said to be singular.

Thus, for a zero operator A, the corresponding matrix will have zero entries. It can also be shown that A is invertible (nonsingular) iff $Ax = 0$ implies $x = 0$.

Let V, W, and A be as before; then for every vector $x \in V$:

$$x = \sum_{j=1}^{n} \xi_j e_j \tag{B.2}$$

where the ξ_j's are scalars, called the "coordinates." Since $Ax \in W$, then:

$$Ax = \sum_{i=1}^{m} \gamma_i f_i \tag{B.3}$$

Now we can claim the following important result via a theorem.

Theorem 1

If we designate Γ and Φ by:

$$\Gamma = \begin{bmatrix} \gamma_1 \\ \gamma_2 \\ \cdot \\ \cdot \\ \cdot \\ \gamma_m \end{bmatrix} \text{ and } \Phi = \begin{bmatrix} \xi_1 \\ \xi_2 \\ \cdot \\ \cdot \\ \cdot \\ \xi_n \end{bmatrix}$$

Then the following is true:

$$\Gamma = M_A \Phi$$

or, equivalently,

$$\gamma_i = \sum_{j=1}^{n} a_{ij} \xi_j, \text{ for } i = 1, 2, \ldots, n$$

Proof

$$Ax = \sum_{j=1}^{n} \xi_j A e_j = \sum_{j=1}^{n} \xi_j \sum_{i=1}^{m} a_{ij} f_i = \sum_{i=1}^{m} \left(\sum_{j=1}^{n} a_{ij} \xi_j \right) f_i$$

where Eq. (B.1) has been used. Now the above equation equated with Eq. (B.3) yields the result.

Definition 7

Let $\{e_1, e_2, \ldots, e_n\}$ and $\{h_1, h_2, \ldots, h_n\}$ be bases in V. Since $h_j \in V$ for all $j = 1, 2, \ldots, n$ and $\{e_1, e_2, \ldots, e_n\}$ is a basis, then

$$h_j = \sum_{i=1}^{n} P_{ij} e_i, \quad j = 1, 2, \ldots, n \tag{B.4}$$

Now the matrix $P = [P_{ij}]_{n \times n}$ is called the matrix of transition from the basis $\{e_1, e_2, \ldots, e_n\}$ to the basis $\{h_1, h_2, \ldots, h_n\}$.

Let Q denote the matrix of transition from $\{h_1, h_2, \ldots, h_n\}$ to $\{e_1, e_2, \ldots, e_n\}$; then it is simple to verify that:

$$Q = P^{-1} \tag{B.5}$$

Let $x \in V$, then

$$x = \sum_{i=1}^{n} \xi_i e_i = \sum_{i=1}^{n} \gamma_i h_i$$

where ξ_i's and γ_i's are coordinates.

It would be very easy to verify that:

$$\xi_i = \sum_{j=1}^{n} P_{ij} \gamma_j \tag{B.6}$$

or, equivalently,

$$\begin{bmatrix} \xi_1 \\ \xi_2 \\ \cdot \\ \cdot \\ \cdot \\ \xi_n \end{bmatrix} = \begin{bmatrix} p_{11} & p_{12} & \cdots & p_{1n} \\ p_{21} & p_{22} & \cdots & p_{2n} \\ \cdot \\ \cdot \\ \cdot \\ p_{n1} & p_{n2} & \cdots & p_{nn} \end{bmatrix} \begin{bmatrix} \gamma_1 \\ \gamma_2 \\ \cdot \\ \cdot \\ \cdot \\ \gamma_n \end{bmatrix} \qquad (B.7)$$

With the above background, we are ready to state a major result in linear algebra given via Theorem 2. The proof will not be given here.

Theorem 2

Let T be a linear transformation from $V \to W$ and let $\{e_1, e_2, \ldots, e_n\}$ and $\{e'_1, e'_2, \ldots, e'_n\}$ be bases in V and let $\{f_1, f_2, \ldots, f_m\}$ and $\{f'_1, f'_2, \ldots, f'_m\}$ be basis in W.

Let $\mathscr{A}$ denote the matrix of T with respect to bases $\{e_1, e_2, \ldots, e_n\}$ and $\{f_1, f_2, \ldots, f_m\}$ and $\mathscr{C}$ be a matrix with respect to the bases $\{e'_1, e'_2, \ldots, e'_n\}$ and $\{f'_1, f'_2, \ldots, f'_m\}$, respectively. Also let S and U denote the matrices of transition from $\{e_1, e_2, \ldots, e_n\}$ to $\{e'_1, e'_2, \ldots, e'_n\}$ and from $\{f_1, f_2, \ldots, f_m\}$ to $\{f'_1, f'_2, \ldots, f'_m\}$, respectively. Then,

$$\mathscr{C} = U^{-1} \mathscr{A} S \qquad (B.8)$$

For Proof see any linear algebra book.

Important Corollary

If $T: V \to V$, then

$$\mathscr{C} = S^{-1} \mathscr{A} S \qquad (B.9)$$

Because $S = U$, its substitution in (B.8) will yield the result.

Definition 8

We now define eigenvalues and eigenvectors, which are used often in our analysis.

Let $A: V \to V$; then if

$$Ax = \lambda x$$

where $x \in V$ and λ is a scalar, then x is called an eigenvector and λ is called the eigenvalue. In general, 0 is an eigenvalue iff $Ax = 0x = 0$ for some $x \neq 0$, i.e., A is singular (not invertible). If $A = I$ (identity operator), then $Ix = x \leftrightarrow \lambda = 1$. In the definition $Ax = \lambda x$, we say x belongs to λ.

Discussion

(1) If $Ax = \lambda x$, then

$$Ax - \lambda x = 0 \leftrightarrow (Ax - \lambda x) = (A - \lambda I)x = 0$$

Thus, x is an eigenvector iff $(A - \lambda I)$ is a singular operator which is equivalent to saying that:

$$\text{determinant } M_{(A-\lambda I)} = |M_{(A-\lambda I)}| = 0$$

(2) From now on, we shall use A for M_A, if there is no confusion about M_A with respect to the specific basis, since there is a 1-1 correspondence and onto mapping from A to M_A (isomorphism).

Definition 9

If a matrix A^* satisfies:

$$A^* = (\bar{a}_{ij})^T$$

where the bar denotes the complex conjugate, and T denotes the transpose, then A^* is said to be an adjoint matrix of A (operator). If $A^* = A$, then A is said to be self-adjoint.

Definition 10

An inner product on vector space V (over the real or complex field) is a complex number such that for every $x, y, z \in V$ and for any scalars α and β the following are satisfied:

(1) $(x, y) = \overline{(y, x)}$

(2) $(\alpha x + \beta y, z) = \alpha(x, z) + \beta(y, z)$

(3) (x, x) iff $x \neq 0$

Definition 11

The norm of a vector x denoted by $\|x\|$ is defined via:

$$\|x\|^2 = (x, x)$$

The following conditions hold and they are given via a proposition.

Proposition 1

(1) $\|x\| > 0$ if $x \neq 0$ and $\|x\| = 0$ iff $x = 0$

(2) $\|\alpha x\| = \alpha \|x\|$ for all scalars α

(3) $\|x + y\| \leq \|x\| + \|y\|$

(4) $\|(x, y)\| \leq \|x\| \|y\|$

Property 4 is the most important and is the only nontrivial one. This property is called the Cauchy-Schwarz Inequality. To prove this relation, let x and y be any given vectors in V, and α and β be any scalars, then we know that:

$$\|\alpha x - \beta y\|^2 \geq 0.$$

However,

$$\|\alpha x - \beta y\|^2 = (\alpha x - \beta y, \alpha x - \beta y) = |\alpha|^2 (x, x) - 2 \operatorname{Re} x \overline{\beta}(x, y) + |\beta|^2 (y, y).$$

Now if we choose $\alpha = (y, y)$ and $\beta = (x, y)$, we get:

$$\|\alpha x - \beta y\|^2 = \|y\|^2 (\|x\|^2 \|y\|^2 - |(x, y)|^2) \geq 0.$$

For the nontrivial case, where $y \neq 0$, the above implies

$$\|x\|^2 \|y\|^2 - |(x, y)|^2 \geq 0$$

which proves the assertion.

By using property 4 it is easy to prove property 3:

$$\|x+y\|^2 = (x+y, x+y) = \|x\|^2 + \text{Re}(x, y) + \|y\|^2$$

$$\leq \|x\|^2 + 2|(x, y)| + \|y\|^2 \leq \|x\|^2 + 2\|x\|\,\|y\| + \|y\|^2$$

$$\leq (\|x\| + \|y\|)^2.$$

which implies $\|x+y\| \leq \|x\| + \|y\|$, as asserted.

Definition 12

A collection of vectors $e_1, e_2, \ldots, e_n$ is orthonormal if $(e_i, e_j) = \delta_{ij}$, where

$$\delta_{ij} = \begin{cases} 1, & \text{if } i = j \\ 0, & \text{if } i \neq j \end{cases}$$

It is left as an exercise to verify that for any collection of orthonormal vectors is a linearly independent set, if $\{e_1, \ldots, e_n\}$ forms a basis, then any $x \in V$ can be given as

$$x = \sum_{i=1}^{n} (x, e_i) e_i \qquad (B.10)$$

This representation is called Bessel's Equality. It can also be verified that:

$$\sum_{i=1}^{n} |(x, e_i)|^2 = \sum_{i=1}^{n} |d_i|^2 \qquad (B.11)$$

where $x = \sum_{i=1}^{n} d_i e_i$. The relation (B.11) is called Parseval's Equality.

An important result concerning orthonormalization is given below.

Theorem 3

Every finite-dimensional vector space V with an inner product defined on it has an orthonormal basis.

Proof

Let $\{x_1, \ldots, x_n\}$ be any basis in V. Consider $f_1, f_2, \ldots, f_n$ defined by

$$f_1 = x_1$$

$$f_2 = x_2 - \alpha x_1, \text{ such that } (f_1, f_2) = 0.$$

$$f_3 = x_3 - \alpha_2 f_2 - \alpha_1 f_1, \text{ such that } (f_1, f_3) = 0, \text{ and } (f_2, f_3) = 0, \text{ etc.}$$

From the above, $\alpha = (x_2, f_1)/\|f_1\|^2$, $\alpha_1 = (x_3, f_1)/\|f_1\|^2$, and $\alpha_2 = (x_3, f_2)/\|f_2\|^2$. Continuing in this manner, we obtain:

$$f_n = x_n - \frac{(x_n, f_{n-1})}{\|f_{n-1}\|^2} f_{n-1} - \frac{(x_n, f_{n-2})}{\|f_{n-2}\|^2} f_{n-2} \cdots - \frac{(x_n, f_1)}{\|f_1\|^2} f_1 \quad (B.12)$$

where, $(f_i, f_j) = 0 \; \forall \; i \neq j$. However, $(f_i, f_j) \neq 1$, in general. Thus, all we need is to normalize each member of the orthogonal set $\{f_1, f_2, \ldots, f_n\}$. This will yield a new set $\{e_1, e_2, \ldots, e_n\}$, where e_i is given by

$$e_i = \frac{f_i}{\|f_i\|}, \text{ for } i = 1, 2, \ldots, n \quad (B.13)$$

Now it becomes obvious that $(e_i, e_j) = \delta_{ij}$. It is easy to verify that (B.13) can take a form given by

$$f_n = x_n - (x_n, e_{n-1})e_{n-1} - (x_n, e_{n-2})e_{n-2} \cdots - (x_n, e_1)e_1. \quad (B.14)$$

We shall now deal with another important aspect of vector spaces: the decomposition of V into subspaces that are orthogonal.

Decomposition of Vector Spaces

Let V be a vector space and V_1 and V_2 be two subspaces, such that $V_1 \cap V_2 = \{0\}$ and $V = V_1 + V_2$, then for every $x \in V$ we can write $x = x_1 + x_2$, where $x_1 \in V_1$ and $x_2 \in V_2$. It is easy to show that if $x = y_1 + y_2$, then $x_1 = y_1$ and $x_2 = y_2$. That is, the representation is unique. The proof is simple. Let $x = x_1 + x_2 = y_1 + y_2$. This is true iff $x_1 - y_1 = y_2 - x_2$. Since x_1 and y_1 are $\in V_1$, then $x_1 - y_1$ is $\in V_1$, but $x_1 - y_1$ is equal to $y_2 - x_2$ which is a member of V_2, thus, $x_1 - y_1 \in V_1 \cap V_2$, which implies $x_1 - y_1 = 0$ or $x_1 = y_1$. Similarly, $x_2 = y_2$ and the proof is done.

Definition 13

If $V = V_1 + V_2$ such that V_1 and V_2 are subspaces and $V_1 \cap V_2 = \{0\}$, then V is said to be the direct sum of V_1 and V_2 and it is denoted by $V = V_1 \oplus V_2$.

Definition 14

Let $x \in V$, and $x = x_1 + x_2$, where $x_1 \in V_1$ and $x_2 \in V_2$. We call x_1 the projection of x onto V_1 along V_2 and x_2 the projection of x onto V_2 along V_1. If P and Q are operators defined such $Px = x_1$ and $Qx = x_2$, we say P and Q are the projections of x as described above.

The projection operators P and Q are linear, and it is easy to verify that:

(1) $P^2 = P$ and $Q^2 = Q$ with $PQ = QP = 0$

(2) $(P + Q) = I$, where I is the identity operator

It is left as an exercise to show that

$$V = R(P) \oplus n(P) \tag{B.15}$$

where $R(P)$ is the range of P and $n(P)$ is the nullity of P. That is, $R(P) = \{P(x) : x \in V\}$ and $n(P) = \{x \in V : P(x) = 0\}$.

Definition 15

If $V = V_1 \oplus V_2$ such that for every $x \in V_1$ and $y \in V_2$, $(x, y) = 0$, then we call V_2 the orthogonal complement of V_1 and denote it by $V_1^\perp$.

Infinite-Dimensional Vector Spaces

A vector space is infinite dimensional if it does not have a basis consisting of a finite number of elements. An infinite-dimensional vector space with an inner product defined on it does not, in general, satisfy $\|x\|^2 = (x, y)$. For infinite-dimensional spaces, the definition of a norm is given via the first three conditions of proposition 1. Any vector space with a norm defined on it, is called a normed vector space. In this book we are interested in problems concerning stochastic processes, where we need to define a special infinite-dimensional vector space called the Hilbert space.

Definition 16

A Hilbert space is a normed vector space that satisfies the following two conditions:

(1) The norm arises from the inner product, i.e., $\|x\| = (x, x)^{1/2}$.

(2) For every sequence of vectors $x_1, x_2, \ldots, x_n, \ldots$ in the normed vector space that converges to x (with respect to the norm), then x must be an element of that space.

Next we shall define a particular Hilbert space. Let S be a sample space over the real or complex scalars such that, for every random variable $X = X(\omega)$, the conditions

$$EX = 0 \quad \text{and} \quad E|X|^2 \leqslant \infty \tag{B.16}$$

are satisfied. We shall also assume that two random variables X_1 and X_2 are equivalent if

$$E|X_1 - X_2|^2 = 0. \tag{B.17}$$

Now let us summarize a major result via the theorem below.

Theorem 4

Let H denote the set of all random variables satisfying the above conditions and let the inner product (X, Y) be given by:

$$(X, Y) = E(X\overline{Y}) \tag{B.18}$$

then H is a Hilbert space with the norm $\|X\|$ defined by:

$$\|X\| = (X, X)^{1/2} = (E|X|^2)^{1/2}. \tag{B.19}$$

The bar denotes the complex conjugate.

The proof is simple. All we need is to show that for any pair of vectors X and $Y \in H$ and scalars α and β

$$\alpha X + \beta Y \in H.$$

Remark. If $X(t)$ is a stochastic process, then this process can be considered as a collection of random variables. That is, for every t, the process $X(t)$ corresponds to a point in H.

We now consider the ramification of the projection theorem concerning the Hilbert space H and discuss a very significant result.

Let $X_1, X_2, \ldots, X_n$ be random variables and M be the vector space which is generated by these variables, i.e., let M be their linear span. Let X be a random variable such that $X \notin H$; then the projection of X onto M along $M^\perp$ is given via:

$$PX = \alpha_1 X_1 + \alpha_2 X_2 + \cdots + \alpha_n X_n \tag{B.20}$$

for scalars $\alpha_i, i = 1, \ldots, n$. Now we shall state the following result via a theorem.

Theorem 5

If $X \in H$ and M is the space generated by random variables $X_1, X_2, \ldots, X_n$ described above, then

$$\left\| X - \sum_{i=1}^{n} \alpha_i X_i \right\|^2 = E \left| X - \sum_{i=1}^{n} \alpha_i X_i \right|^2 \tag{B.21}$$

is minimum iff

$$E\left[\left(X - \sum_{i=1}^{n} \alpha_i X_i \right) \overline{X}_j \right] = 0, \quad \text{for } j = 1, 2, \ldots, n. \tag{B.22}$$

Note that the above would be true iff $\sum_{i=1}^{n} \alpha_i X_i$ is the orthogonal projection of X, i.e., $P(X) = \sum_{i=1}^{n} \alpha_i X_i$. The proof of this theorem is given in Chapter 4 via Theorem 2.

Definition 17

If $A = A^*$, then (Ax, x) is said to be positive definite if

$$(Ax, x) > 0, \text{ for all } x \neq 0$$

and negative definite if

$$(Ax, x) < 0, \text{ for all } x \neq 0$$

Similarly, if A satisfies

$$(Ax, x) \geq 0, \text{ for all } x \neq 0$$

then A is said to be positive semi-definite; the definition of negative semi-definiteness is done in a similar manner.

Definition 13

If A is none of the above, A is said to be indefinite, that is, $(Ax, x) > 0$ for some x and $(Ax, x) < 0$ for another x.

Definition 14

The quadratic form of $A = A^*$ is defined via

$$Q(x) = \sum_{i=1}^{n} \sum_{j=1}^{n} a_{ij} \xi_i \xi_j^*$$

where ξ_i's are the coordinates of the vector x.

The above background should suffice to support the material in the text.

APPENDIX C

FOURIER AND BILATERAL LAPLACE TRANSFORMS AND THEIR INVERSIONS

The power spectrum is the Fourier transform of the wide-sense stationary autocorrelation function. Thus, the manipulation of the Fourier transform and its corresponding inverse is extremely important. If a function $f(t)$ has a Fourier transform, it will also have a bilateral Laplace transform. The inverse of each transform is unique; however, it is easier to obtain the inverse of a bilateral Laplace transform. Thus, the procedure of obtaining the inverse Fourier transform is to obtain the corresponding bilateral Laplace transform and apply the inversion formula. Thus, in what follows, a discussion of Fourier and bilateral Laplace transform is made.

Before we get involved with the concepts, we need some mathematical tools such as definitions and theorems; however the proofs are not provided.

Definition 1

A function (complex) $f(s)$ is analytic at s_0 if f is single valued and differentiable at s_0.

Theorem 1 (Cauchy's Integral Theorem)

Given the function $f(s)$ such that f is analytic at all points within and on any closed curve C in the complex plane, then

$$\int_C f(s)\,ds = 0$$

where the integral designates the integral along the closed path C.

Theorem 2 (Cauchy's Integral Formula)

Let f and C be as above; then for any point a which is an interior point in C, the following is true:

$$f(a) = \frac{1}{2\pi j} \int_C \frac{f(s)\,ds}{s - a} \tag{C.1}$$

The result is proven via the aid of Theorem 1. Thus, in Theorem 2 every analytic function $f(s)$ is completely determined in the interior of a given close curve C, where the values of $f(s)$ are given on C only. Next the last two theorems are extended to get an important result which we shall give via Theorem 3, but first the singularities.

Definition 2

If $f(s)$ is not analytic at point s_0, then s_0 is called a singular point. If there is a neighborhood of $s = s_0$ such that $f(s)$ has no other singular point, then s_0 is called the isolated singularity and, unless specified otherwise, all the singularities in the appendix are isolated singularities.

Example

$f(s) = 1/s$ has an isolated singularity at $s = 0$, since the neighborhood given by $|s| = \rho > 0$ contains no singularity other than 0. Similarly,

$$f(s) = \frac{s-1}{s(s^2+4)}$$

has three isolated singularities at $s = 0$, $s = 2j$, $s = -2j$. The function

$$f(s) = \exp\left\{\frac{1}{1-s^2}\right\}$$

has two isolated singular points at $s = 1$ and $s = -1$.

Note that in the first two cases, the singularities are poles, and in the third case it is not a pole.

Another Example

The function

$$f(s) = \frac{1}{\sin\left(\frac{1}{s}\right)}$$

has singularities at $s = \pm 1/(k\pi)$, $k = 1,2,\ldots$. These singularities are isolated; however at $s = 0$, the singularity is not isolated, regardless of how small the radius ρ of the circle $|s| = \rho$ may be.

If $f(s)$ has an isolated singularity at $s = s_0$, then $f(s)$ can be represented via the infinite series:

$$f(s) = b_0 + b_1(s - s_0) + b_2(s - s_0)^2 + \cdots + \frac{b_{-1}}{s - s_0} + \frac{b_{-2}}{(s - s_0)^2} + \cdots$$

$$= \sum_{n=0}^{\infty} b_n (s - s_0)^n + \sum_{n=1}^{\infty} \frac{b_{-n}}{(s - s_0)^n} \qquad (C.2)$$

Definition 3

The above series is called Laurent's series and b_{-1} is called the residue of $f(s)$ at the singularity $s = s_0$.

Definition 4

A special case is where

$$f(s) = \sum_{n=0}^{\infty} b_n (s - s_0)^n + \frac{b_{-1}}{s - s_0} + \frac{b_{-2}}{(s - s_0)^2} + \cdots + \frac{b_{-m}}{(s - s_0)^m} \qquad (C.3)$$

The singularity (isolated) $s = s_0$ is called a pole of order m.

Remark. For Eq. (C.3) b_{-1} is given by:

$$b_{-1} = \frac{1}{(m - 1)!} \left. \frac{d^{m-1}[(s - s_0)^m f(s)]}{ds^{m-1}} \right|_{s=s_0} \qquad (C.4)$$

If $m = 1$, then $s = s_0$ is said to be a simple pole and (C.4) reduces to:

$$b_{-1} = \lim_{s \to s_0} f(s)(s - s_0) \tag{C.5}$$

Theorem 3

Let $f(s)$ be analytic in the given region R bounded by the closed curve C and let $s_1, s_2, \ldots, s_m$ be the isolated singularities of $f(s)$ in the interior of C, then

$$\int_C f(s)\, ds = 2\pi j \sum_{k=1}^{m} (b_{-1})_k \tag{C.6}$$

where $(b_{-1})_k$ is the residue corresponding to s_k.

The result is called the residue theorem which states that regardless of how complicated the calculation of integral of $f(s)$ around the contour is, it can be obtained by the summation of all residues multiplied by $2\pi j$.

Equation (C.6) will play a major role in the inversion process of a transform.

Definition 5

Let $f(t)$ and $F_B(s)$ be functions defined by:

$$F_B(s) = \int_{-\infty}^{\infty} f(t) \exp(-st)\, dt \tag{C.7}$$

Then we say $F_B(s)$ is the bilateral Laplace transform of $f(t)$, provided that $F_B(s)$ exists in some region $\sigma_1 < \sigma < \sigma_2$.

Theorem 4

If $F_B(s)$ exists, then $f(t)$ can be obtained:

$$f(t) = \frac{1}{2\pi j} \lim_{R \to \infty} \int_{d-j}^{d+j} F_B(s) \exp(st)\, ds \tag{C.8}$$

where d and R are given via the sketch and $\sigma_1 < d < \sigma_2$ (see sketch).

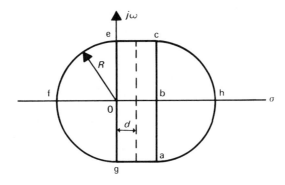

Proof

For the bilateral transform, the regions of convergence for $f(t)$ is generally given by $\sigma_1 < \sigma < \sigma_2$. However, for the one-sided Laplace transform, the region of convergence is normally given by $\sigma > \sigma_0$.

For $t > 0$, then we can show:

$$\lim_{R \to \infty} \int_{cefga} F_B(s) \exp(st)\, ds = 0 \tag{C.9}$$

and for $t < 0$:

$$\lim_{R \to \infty} \int_{cha} F_B(s) \exp(st)\, ds = 0 \tag{C.10}$$

Equations (C.9) or (C.10), together with (C.8), implies that abc may be changed to abcefg for $t < 0$ and, for $t > 0$, abc can be changed to abcha. However, either abcefg or abcha is a closed contour enclosing all the singularities as long as $R \to \infty$, which implies we can directly use the residue theorem (Theorem 3).

Thus, given $t > 0$,

$$f(t) = \frac{1}{2\pi j} \int_{abcha} F_B(s) \exp(st)\, ds = \sum (b_{-1})_k \qquad \text{(C.11)}$$

where $(b_{-1})_k$ is the residue of the kth singularity to the left of abc. For $t < 0$, $f(t)$ is given by:

$$f(t) = \frac{1}{2\pi j} \int_{abcha} F_B(s) \exp(st)\, ds = -\sum (b_{-1})_k \qquad \text{(C.12)}$$

where $(b_{-1})_k$ is the residue of the kth singularity to the right of abc. The negative sign signifies the fact that the direction of abcha is clockwise and, therefore, negative. Thus, we have proven the inversion formula.

If $f(t)$ is absolutely integrable, i.e.,

$$\int_{-\infty}^{\infty} |f(t)|\, dt < \infty$$

then we shall define

$$\mathscr{F}(\omega) = \int_{-\infty}^{\infty} f(t) \exp(-j\omega t)\, dt \qquad \text{(C.13)}$$

as the Fourier transform of $f(t)$. It can be shown that given $\mathscr{F}(\omega)$, $f(t)$ satisfies:

$$f(t) = \frac{1}{2\pi} \int_{-\infty}^{\infty} \mathscr{F}(\omega) \exp(j\omega t)\, d\omega \qquad \text{(C.14)}$$

Basic Fourier Transform Pairs

$f(t)$	$G(\nu) = \mathscr{F}(2\pi\nu), \ \omega = 2\pi\nu$
(1) $\delta(t)$	1
(2) 1	$\delta(\nu)$
(3) $\cos \omega_0 t$	$\dfrac{1}{2}[\delta(\nu - \nu_0) + \delta(\nu + \nu_0)]$
(4) $\sin \omega_0 t$	$\dfrac{1}{2}[\delta(\nu - \nu_0) - \delta(\nu + \nu_0)]$
(5) $\dfrac{\sin(2\pi Wt)}{(2\pi Wt)}$	rectangular pulse of height $\dfrac{1}{2W}$ from $-W$ to W
(6) rectangular pulse of height $\dfrac{1}{2T}$ from $-T$ to T	$\dfrac{\sin(2\pi T\nu)}{(2\pi T\nu)}$
(7) $\left(\dfrac{\sin 2\pi Wt}{2\pi Wt}\right)^2$	triangular pulse of height $\dfrac{1}{2W}$ from $-2W$ to $2W$
(8) $\exp(j\omega_0 t)$	$\delta(\nu - \nu_0)$
(9) $\delta(t - \tau)$	$\exp(-j\omega\tau)$
(10) $\begin{cases} \dfrac{1}{\tau}\exp(-t/\tau), & t \geq 0 \\ 0, & t < 0 \end{cases}$	$\dfrac{1}{1 + j\omega\tau}$
(11) $\exp(-|\alpha|t), \ \alpha > 0$	$\dfrac{2\alpha}{\omega^2 + \alpha^2}$
(12) $\dfrac{1}{\sqrt{2\pi}\,\sigma}\exp\left\{\dfrac{-(t-m)^2}{2\sigma^2}\right\}$	$\exp\left(-j\omega m - \dfrac{\omega^2\sigma^2}{2}\right)$

Equations (C.13) and (C.14) are called the Fourier transform pair. Now if the Fourier transform of $f(t)$ exists, then for a fixed $\sigma > 0$, the Fourier transform of $f(t) \exp(-\sigma t)$ would also exist (it is absolutely integrable). Then

$$\int_{-\infty}^{\infty} [f(t) \exp(-\sigma t)] \exp(-j\omega t) \, dt = \int_{-\infty}^{\infty} f(t) \exp[-(\sigma + j\omega t)] \, dt$$

Let $s = \sigma + j\omega$ and denote the right-hand-side of the integral as $F(\sigma + j\omega)$ or $F(s)$. Now it is obvious that the function $f(t) \exp(-\sigma t)$, given its Fourier transform $F(\sigma + j\omega)$, is:

$$f(t) \exp(-\sigma t) = \mathscr{F}^{-1}[F(\sigma + j\omega)] = \frac{1}{2\pi} \int_{-\infty}^{\infty} F(\sigma + j\omega) \exp(j\omega t) \, d\omega$$

The last equation utilizes the inversion formula of a Fourier transform. Multiplying both sides of the equation by $\exp(\sigma t)$, we get:

$$f(t) = \frac{1}{2\pi} \int_{-\infty}^{\infty} F(\sigma + j\omega) \exp(\sigma + j\omega) t \, d\omega$$

Now making the change of variable $s = \sigma + j\omega$ will yield:

$$f(t) = \frac{1}{2\pi j} \int_{-\infty}^{\infty} F(s) \exp(st) \, ds \qquad (C.15)$$

However, $F(s)$ is exactly the bilateral transform $F_B(s)$. Thus, we shall utilize the bilateral Laplace transform inversion formula.

Note that the inversion of both Fourier and bilateral transforms are unique and if the Fourier transform of a waveform $f(t)$ exists, so does its bilateral transform. The bilateral transform $F_B(s)$ can be obtained from $\mathscr{F}(j\omega)$ in a unique manner, by substituting:

$$F_B(s) = \mathscr{F}(j\omega)\big|_{s=j\omega}$$

APPENDIX D

A SPECIAL VECTOR SPACE

Let V_N be an N-dimensional vector space over a complex field. Let $\{f_1, f_2, \ldots, f_N\} = \{f_i\}_{i=1}^{N}$ be any basis in V_N. If there is an inner product defined with an associated norm, then it is a standard result that $\{f_1, f_2, \ldots, f_N\}$ can be orthonormalized. That is, $\{e_i\}_{i=1}^{N}$ is a basis such that:

$$(e_i, e_j) = \begin{cases} 1, & \text{if } i = j \\ 0, & \text{if } i \neq j \end{cases}$$

Now for any vector $x \in V_N$, it can be shown that:

$$x = \sum_{i=1}^{N} (x, e_i) e_i$$

and

$$\|x\|^2 = \sum_{i=1}^{N} |(x, e_i)|^2$$

The idea of orthonormalization can be extended to the infinite dimensional case.

An Infinite Dimensional Vector Space

Let L_2 denote the set of all piecewise continuous functions over $[0,2\pi]$ such that:

$$\int_0^{2\pi} |f(t)|^2 \, dt < \infty \tag{D.1}$$

It can be verified that L_2 is a vector space under the usual operations of functions: $(f+g)(t) = f(t) + g(t)$ and $(\alpha f)(t) = \alpha(f(t))$.

Now let us define the inner product (f,g) by:

$$(f,g) = \int_0^{2\pi} f(t)\, \overline{g}(t) \, dt \tag{D.2}$$

where the bar denotes the conjugate. Thus, the corresponding norm is given by:

$$\|f\|^2 = (f,f) = \int_0^{2\pi} |f(t)|^2 \, dt \tag{D.3}$$

A simple computation shows that $\exp(jnt)$ for $n = 0, \pm 1, \pm 2, \ldots$ are mutually orthogonal in L_2 and it can be shown that:

$$(\exp(jmt), \exp(jnt)) = \begin{cases} 0, & \text{if } m \neq n \\ 2\pi, & \text{if } m = n \end{cases} \tag{D.4}$$

However, we can orthonormalize the collection

$$\{\exp(jnt)\}_{n=-\infty}^{n=\infty}$$

by letting

$$e_n(t) = \frac{1}{\sqrt{2\pi}} \exp\{jnt\}$$

L_2 with an orthonormal basis is said to be a complete space. Recall any finite-dimensional vector space is complete.

Let H be a subspace of L_2 which is generated by

$$\{e_n(t)\}_{n=-\infty}^{n=\infty}$$

that is, H consists of all linear combinations of the form

$$\sum_{n=-\infty}^{n=\infty} \alpha_n e_n$$

where α_n's are scalars.

Now for every $f \in H$, we can write:

$$f(t) = \sum_{n=-\infty}^{\infty} \alpha_n e_n(t) \tag{D.5}$$

where $\alpha_n = (f, e_n)$, and α_n can be written as:

$$\alpha_n = (f, e_n) = \frac{1}{\sqrt{2\pi}} \int_0^{2\pi} f(t) \exp\{-jnt\} dt \tag{D.6}$$

Thus, from Eqs. (D.3) and (D.5), it is easy to verify that:

$$\|f\|^2 = \sum_{n=-\infty}^{\infty} |(f, e_n)|^2 \tag{D.7}$$

Now remembering that:

$$\|f\|^2 = \int_0^{2\pi} |f(t)|^2 \, dt$$

and utilizing the fact that $\alpha_n = (f, e_n)$, we can rewrite:

$$\int_0^{2\pi} |f(t)|^2 \, dt = \sum_{n=-\infty}^{\infty} |\alpha_n|^2 \qquad (D.8)$$

Equation (D.8) is called Parseval's equality.

Important Remarks

(1) It must be emphasized that the expansion

$$f(t) = \sum_{n=-\infty}^{\infty} \alpha_n e_n(t) = \frac{1}{\sqrt{2\pi}} \sum_{n=-\infty}^{\infty} \alpha_n \exp\{jnt\} \qquad (D.9)$$

is not interpreted as saying the series is pointwise converging to the function. Equation (D.9) actually means that $f_n \in L_2$ is given by:

$$f_n(t) = \frac{1}{\sqrt{2\pi}} \sum_{k=-n}^{n} \alpha_k \exp\{jkt\}$$

and converges to f in the norm specified in L_2. That is:

$$\|f - f_n\| = \left[\int_0^{2\pi} |f(t) - f_n(t)|^2 \, dt \right]^{1/2} \underset{n \to \infty}{\to} 0 \qquad (D.10)$$

(2) If we change 2π to T and the interval $[0, 2\pi]$ is changed to $[-T/2, T/2]$, we can then write:

$$f(t) = \sum_{n=-\infty}^{\infty} c_n \exp\{jn\omega_0 t\} \qquad (D.11)$$

where $\omega_0 = 2\pi/T$. Since

$$\{\exp\{jn\omega_0 t\}\}_{n=-\infty}^{\infty}$$

are pairwise orthogonal,

$$(\exp\{jm\omega_0 t\}, \exp\{jn\omega_0 t\}) = \begin{cases} 0, & \text{if } m \neq n \\ T, & \text{if } m = n \end{cases}$$

Thus, we have:

$$f(t) = \sum_{n=-\infty}^{\infty} \alpha_n e_n(t) = \sum_{n=-\infty}^{\infty} \sqrt{T} c_n e_n(t) \qquad (D.12)$$

where $e_n = (1/\sqrt{T}) \exp\{n\omega_0 t\}$ and $\alpha = (f, e_n)$.

Parseval's equality becomes:

$$\|f\|^2 = \int_{-T/2}^{T/2} |f(t)|^2 \, dt = \sum_{n=-\infty}^{\infty} |\alpha_n|^2$$

$$= \sum_{n=-\infty}^{\infty} |\sqrt{T} c_n|^2 = T \sum_{n=-\infty}^{\infty} |c_n|^2$$

From which, we obtain:

$$\frac{1}{T} \int_{-T/2}^{T/2} |f(t)|^2 \, dt = \sum_{n=-\infty}^{\infty} |c_n|^2 \qquad (D.13)$$

The last equation is another form of Parseval's equality.

APPENDIX E
STATE VARIABLES

Let $X(t)$ be an n-vector such that:

$$\dot{X} = A(t) X(t), \quad X(t_0) = X_0 \qquad (E.1)$$

where $X(t)$ and $A(t)$ are continuously differentiable and $A(t)$ is an $n \times n$ matrix. The solution of Eq. (E.1) is given by:

$$X(t) = \phi(t, t_0) X(t_0) = \phi(t, t_0) X_0 \qquad (E.2a)$$

where

$$\dot{\phi} = A(t) \phi(t, t_0), \phi(t_0, t_0) = I \qquad (E.2b)$$

This is easy to verify, since the solution of the differential equation for a specified condition is unique and $X(t)$ in Eq. (E.2) will be a solution with the initial condition:

$$X(t_0) = \phi(t_0, t_0) X_0 = I X_0 = X_0$$

Two Important Properties

Let t_1 and t_2 be two different times such that t_1 and t_2 are $\geq t_0$. Then we have:

$$X(t_2) = \phi(t_2, t_0) X_0 \tag{E.3}$$

and

$$X(t_1) = \phi(t_1, t_0) X_0 \tag{E.4}$$

Now if the initial condition is at t_1, then $X(t_2)$ is given by:

$$X(t_2) = \phi(t_2, t_1) X(t_1) \tag{E.5}$$

Substituting $X(t_1)$ from Eq. (E.4) into Eq. (E.5) yields:

$$X(t_2) = \phi(t_2, t_1) \phi(t_1, t_0) X_0 \tag{E.6}$$

Comparing (E.3) and (E.6) gives rise to:

$$\phi(t_2, t_0) = \phi(t_2, t_1) \phi(t_1, t_0) \tag{E.7}$$

As a special case of Eq. (E.7), let $t_2 = t_0$. Then

$$\phi(t_0, t_0) = I = \phi(t_0, t_1) \phi(t_1, t_0)$$

from which

$$\phi(t_1, t_0)^{-1} = \phi(t_0, t_1) \tag{E.8}$$

Equations (E.7) and (E.8) are very important. It can be verified that $\phi(\cdot,\cdot) \neq 0$. From Eq. (E.8) it is obvious that the inverse of $\phi(t_1, t_0)$ is obtained by changing the arguments t_1 and t_0 to t_0 and t_1, respectively.

Example 1

From

$$\dot{X} = 2X, \quad X(t_0) = X_0$$

Solve the differential equation via the transition matrix.

$$\dot{\phi} = 2\phi, \quad \phi(t_0, t_0) = 1$$

will imply that

$$\phi(t, t_0) = \exp\{2(t - t_0)\}$$

Thus,

$$X(t) = \phi(t, t_0) X_0 = X_0 \exp\{2(t - t_0)\}$$

Example 2

Repeat Example 1 for:

$$\dot{X} = a(t) X(t), \quad X(t_0) = X_0$$

Solution

$$\dot{\phi} = a(t) \phi(t, t_0), \quad \phi(t_0, t_0) = 1$$

implies that $\dot{\phi}/\phi = a(t)$ from which we get:

$$\phi(t, t_0) = \exp\left\{\int_{t_0}^{t} a(t)\, dt\right\}$$

Thus,

$$X(t) = X_0 \exp\left\{\int_{t_0}^{t} a(t)\, dt\right\}$$

General Solution with Forcing Function Inputs

Consider the general time-varying differential equation:

$$\dot{X} = A(t)\, X(t) + B(t)\, U(t) \qquad \text{(E.9a)}$$

$$Y(t) = C(t)\, X(t) + D(t)\, U(t) \qquad \text{(E.9b)}$$

Assume the solution $X(t)$ exists and

$$X(t_0) = X_0$$

is the initial condition. We claim $X(t)$ is given by:

$$X(t) = \phi(t, t_0)\, X_0 + \int_{t_0}^{t} \phi(t, t_0)\, \phi^{-1}(\lambda, t_0)\, B(\lambda)\, U(\lambda)\, d\lambda \qquad \text{(E.10a)}$$

$$= \phi(t, t_0)\, X_0 + \int_{t_0}^{t} \phi(t, \lambda)\, B(\lambda)\, U(\lambda)\, d\lambda \qquad \text{(E.10b)}$$

Let us verify Eq. (E.10). For convenience, we shall not write the arguments in t. Let

$$X(t) \triangleq \phi(t, t_0)\, Z(t) \text{ or, equivalently, } Z(t) \triangleq \phi^{-1}(t, t_0)\, X(t) \qquad \text{(E.11)}$$

Taking the derivative of both sides yields:

$$\dot{X} = \dot{\phi} Z + \phi \dot{Z} \qquad \text{(E.12)}$$

Equating the right-hand side of Eq. (E.9a) with (E.12) gives rise to:

$$AX + BU = \dot{\phi}Z + \phi\dot{Z} \tag{E.13}$$

Now from $\dot{\phi} = A\phi$ we assert that:

$$\dot{\phi}Z = A\phi Z = AX \tag{E.14}$$

where in the above we have used Eq. (E.11).

Substituting (E.14) into Eq. (E.12) yields:

$$\phi\dot{Z} = BU \text{ or, equivalently, } \dot{Z} = \phi^{-1} BU \tag{E.15}$$

where upon integration, we get:

$$Z(t) = Z(t_0) + \int_{t_0}^{t} \phi^{-1}(\lambda, t_0) B(\lambda) U(\lambda) \, d\lambda \tag{E.16}$$

Utilizing Eq. (E.11) and the fact that $Z(t_0) = X(t_0)$, we obtain:

$$X(t) = \phi(t, t_0) X_0 + \int_{t_0}^{t} \phi(t, t_0) \phi^{-1}(\lambda, t_0) B(\lambda) U(\lambda) \, d\lambda \tag{E.17}$$

which concludes the first part of the proof.

To prove the second part, we make use of $\phi^{-1}(\lambda, t_0) = \phi(t_0, \lambda)$ which implies:

$$\phi(t, t_0) \phi^{-1}(\lambda, t_0) = \phi(t, t_0) \phi(t_0, \lambda) = \phi(t, \lambda) \tag{E.18}$$

Substituting (E.18) into (E.17) gives rise to:

$$X(t) = \phi(t, t_0) X_0 + \int_{t_0}^{t} \phi(t, \lambda) B(\lambda) U(\lambda) \, d\lambda \tag{E.19}$$

which is the desired result.

Substituting (E.17) or (E.19) into (E.9b) will yield the output. Therefore,

$$Y(t) = C(t) X(t) + D(t) U(t)$$

$$= C(t) \left[\phi(t, t_0) X_0 + \int_{t_0}^{t} \phi(t, \lambda) B(\lambda) U(\lambda) d\lambda \right] + D(t) U(t)$$

(E.20)

Thus, the most important part of the solution is acquisition of the transition matrix $\phi(\cdot,\cdot)$, which is needed to solve $X(t)$. Once $X(t)$ is known, $Y(t)$ can be obtained immediately (see E.20).

To obtain $\phi(\cdot,\cdot)$ for the time-varying case is not easy and the general equation

$$\dot{\phi} = A(t) \phi(t, t_0), \quad \phi(t_0, t_0) = I$$

must be solved for. However, for the time-invariant case, where A, B, C, and D are constant matrices, the solution is considerably easier. Before discussing this special case, let us first define:

$$\exp(At) \triangleq I + At + \frac{A^2 t^2}{2!} + \cdots + \frac{A^n t^n}{n!} + \cdots \qquad (E.21)$$

Now for the time invariant case, $\phi(t, t_0)$ becomes:

$$\phi(t, t_0) = \exp\{A(t - t_0)\} \qquad (E.22)$$

To verify (E.22) is very simple since

$$\frac{d}{dt} \exp\{A(t - t_0)\} = A \exp\{A(t - t_0)\}$$

with $\phi(t_0, t_0) = A^0 = I$. Now, without any loss of generality, assume $t_0 = 0$ and let us state the following claim.

The transition matrix exp $\{At\}$ is obtained as:

$$\exp\{At\} = \mathscr{L}^{-1}(sI - A)^{-1} \tag{E.23}$$

Thus, exp $\{At\}$ is the inverse Laplace transform of $(sI - A)^{-1}$.

The proof is simple. Take the Laplace transform of (E.9a) to get:

$$s\mathscr{X}(s) - X_0 = A\mathscr{X}(s) + B\mathscr{U}(s) \tag{E.24}$$

where $\mathscr{X}(s)$ and $\mathscr{U}(s)$ are corresponding Laplace transforms of $X(\cdot)$ and $U(\cdot)$. This can be done since A and B are both constant matrices. From (E.24), we can get:

$$\mathscr{X}(s) = (sI - A)^{-1} X_0 + (sI - A)^{-1} B\mathscr{U}(s) \tag{E.25}$$

Taking the inverse Laplace transform of the above and equating the result with the right-hand side of (E.19) with $t_0 = 0$, we obtain:

$$\exp\{At\} = \mathscr{L}^{-1}(sI - A)^{-1}$$

as asserted.

REFERENCES

CHAPTER 1

1. A. Papoulis, *Probability, Random Variables, and Stochastic Processes*, McGraw-Hill, 1965.
2. P. Meyer, *Introduction to Probability and Statistical Applications*, Addison Wesley, 1965.
3. E. Wong, *Stochastic Processes in Information and Dynamical Systems*, McGraw-Hill, 1971.
4. H. Cramer and H. Leadbetter, *Stationary and Related Stochastic Processes*, Wiley, 1967.
5. B. Gnedenko, *The Theory of Probability*, Chelsea, 1967.
6. A. Whalen, *Detection of Signals in Noise*, Academic Press, 1971.
7. A. Granino, *Random Processes Simulation and Measurements*, McGraw-Hill, 1966.
8. M. Eisen, *Introduction to Mathematical Probability Theory*, Prentice-Hall, 1969.
9. A. Rényi, *Probability Theory*, North-Holland, 1970.
10. W. Davenport and W. Root, *An Introduction to the Theory of Random Signals and Noise*, McGraw-Hill, 1958.
11. W. Feller, *Probability Theory and Its Applications*, Vol. 1, Wiley, 1950.

CHAPTER 2

1. H. Cramer and H. Leadbetter, *Stationary and Related Stochastic Processes*, Wiley, 1967.

2. B. Lathi, *An Introduction to Random Signals and Communication Theory*, International Textbook Co., 1968.

3. E. Wong, *Stochastic Processes in Information and Dynamical Systems*, McGraw-Hill, 1971.

4. A. Papoulis, *Probability, Random Variables, and Stochastic Processes*, McGraw-Hill, 1965.

5. Y. Rozanov, *Stationary Random Processes*, Holden-Day, 1967.

6. A. Yaglom, *An Introduction to the Theory of Stationary Random Functions*, Prentice-Hall, 1962.

7. E. Parzen, *Modern Probability Theory*, Wiley, 1960.

8. R. Schwarz and B. Friedland, *Linear Systems*, McGraw-Hill, 1965.

9. B. Gnedenko, *The Theory of Probability*, Chelsea, 1967.

10. Y. Khintchine, *Mathematical Methods in the Theory of Queueing*, Griffin, 1960.

11. J. Doob, *Stochastic Processes*, Wiley, 1963.

12. T. McGarty, *Stochastic Systems and State Estimation*, Wiley, 1974.

13. L. Schwarts, *Theorie des Distributions*, Vols. I and II, Herman et Cie., Paris, 1950.

CHAPTER 3

1. C. Chen, *Introduction to Linear System Theory*, Holt, Rinehart and Winston, 1970.

2. K. Ogata, *State Space Analysis of Control Systems*, Prentice-Hall, 1967.

3. L. Zadeh and C. Desoer, *Linear System Theory*, McGraw-Hill, 1963.

4. A. Papoulis, *Probability, Random Variables, and Stochastic Processes*, McGraw-Hill, 1965.

5. N. Nahi, *Estimation Theory and Its Applications*, Wiley, 1969.

6. R. Schwarz and B. Friedland, *Linear Systems*, McGraw-Hill, 1965.

7. B. Lathi, *An Introduction to Random Signals and Communication Theory*, International Textbook Co., 1968.

8. E. Wong, *Stochastic Processes in Information and Dynamical Systems*, McGraw-Hill, 1971.

9. L. Franks, *Signal Theory*, Prentice-Hall, 1969.

10. T. McGarty, *Stochastic Systems and State Estimation*, Wiley, 1974.

CHAPTER 4

1. N. Weiner, *The Extrapolation, Interpolation, and Smoothing of Stationary Time Series with Engineering Applications*, Wiley, 1949.
2. R. Kalman, A New Approach to Linear Filtering and Prediction Problems, *Trans. ASME, J. Basic Eng.*, Vol. 82D, pp. 34-45, March 1960a.
3. R. Kalman, New Methods in Wiener Filtering Theory, *Proc. First Symp. Eng. Appl. Random Functions Theory Probability*, Wiley, 1963.
4. R. Kalman and R. Bucy, New Results in Linear Filtering and Prediction Theory, *Trans. ASME, J. Basic Eng.*, Vol. 83D, pp. 95-108, March 1961.
5. K. Ogata, *State Space Analysis of Control Systems*, Prentice-Hall, 1967.
6. C. Chen, *Introduction to Linear System Theory*, Holt, Rinehart and Winston, 1970.
7. R. Bucy and P. Joseph, *Filtering for Stochastic Processes with Applications to Guidance*, Interscience Publishers, 1968.
8. N. Nahi, *Estimation Theory and Its Applications*, Wiley, 1969.
9. E. Wong, *Stochastic Processes in Information and Dynamical Systems*, McGraw-Hill, 1971.
10. L. Rhodes, A Tutorial Introduction to Estimation and Filtering, *IEEE Trans. Control Theory*, Vol. AC-16, pp. 688-706, Dec. 1971.
11. P. Liebelt, *An Introduction to Optimal Estimation*, Addison Wesley, 1969.
12. Deutsch, *Estimation Theory*, Prentice-Hall, 1965.
13. B. Anderson and J. Moore, The Kalman-Bucy Filter as a True Time Varying Weiner Filter, *IEEE Trans. Syst., Man, Cybernetics*, Vol. SMC-1, pp. 119-128 Apr. 1971.
14. A. Sage and P. Melsa, *Estimation Theory with Applications to Communications and Controls*, McGraw-Hill, 1971.
15. A. Sage, *Optimum Systems Controls*, Prentice-Hall, 1968.
16. E. Dynkin, *Markov Processes*, Springer-Verlag, 1965.
17. W. Fleming, Some Markovian Optimization Problems, *J. Math. Mech.*, Vol. 12, pp. 131-140, 1963.
18. A. Jazwinski, *Nonlinear Filtering Theory*, Academic Press, 1969.
19. H. Kushner, *Stochastic Stability and Control*, Academic Press, 1967.
20. H. Kushner, *Introduction to Stochastic Control*, Holt, Rinehart and Winston, 1970.

21. J. Meditch, *Stochastic Optimal Linear Estimation and Control*, McGraw-Hill, 1969.

22. T. McGarty, *Stochastic Processes and State Estimation*, Wiley, 1974.

23. D. Fraser, On the Applications of Optimum Linear Smoothing Techniques to Linear and Nonlinear Dynamics Systems, Ph.D. Thesis, M.I.T., 1967.

24. T. Kailath, An Innovation Approach to Least Square Estimation – Part I: Linear Filtering in Additive White Noise, *IEEE Trans. Autom. Control*, Vol. AC-13, Dec. 1968.

25. T. Kailath and P. Frost, An Innovation Approach to Least Square Estimation – Part II: Linear Smoothing in Additive White Noise, *IEEE Trans. Autom. Control*, Vol. AC-13, pp. 655–660, Dec. 1968.

26. P. Frost and T. Kailath, An Innovative Approach to Least Square Estimation – Part III: Nonlinear Estimation in White Gaussian Noise, *IEEE Trans. Autom. Control*, Vol. AC-16, pp. 217–226, June 1971.

27. P. Dryer and S. McReynolds, Extention of Square-Root Filtering to Include Process Noise, *J. Optimiz. Theory Appl.*, Vol. 3, pp. 444–459, 1969.

28. R. Battin, *Astronomical Guidance*, McGraw-Hill, 1964.

29. H. Cox, On the Estimation of State Variables and Parameters for Noisy Dynamic Systems, *IEEE Trans. Autom. Control*, Vol. AC-9, pp. 5–12, Jan. 1964.

30. T. Kailath, A View of Three Decades of Linear Filtering Theory, *IEEE Trans. Inform. Theory*, Vol. IT-20, March 2, 1964.

31. A. Bryson and D. Johansen, Linear Filtering for Time Varying Systems, *IEEE Trans. Autom. Control*, Vol. AC-10, pp. 4–10, Jan. 1965.

32. L. Zachrisson, On Optimal Smoothing on Continuous-Time Kalman Processes, *Inform. Sci.*, Vol. 1, pp. 143–172, 1969.

33. D. Fraser and J. Potter, the Optimum Linear Smoother as a Combination of Two Optimum Linear Filters, *IEEE Trans. Autom. Control*, pp. 387–390, Aug. 1969.

34. A. Johnston, T. Assefi, and J. Lai, Automated Guidance Using Discrete Markers, *IEEE Trans. Veh. Technol.*, Vol. VT-28, pp. 95–106, Feb. 1979.

35. T. Assefi and J. Alexander, *Viking Orbiter 75 In-Flight Pointing Calibration of the High Gain Antenna*, to be published.

36. R. Kalman, On General Theory of Control Systems, *Proc. First Intern. Congr. Automatic Control*, pp. 481–493, Butterworth, 1960.

37. B. Anderson and T. Kailath, The Choice of Signal Process Models in Kalman-Bucy Filtering, *J. Math. Appl.*, Vol. 35, pp. 659–668. Sept. 1971.

38. E. Wong, Recent Progress in Stochastic Processes — A Survey, *IEEE Trans. Inform. Theory*, Vol. IT-19, pp. 262–275, May 1973.

CHAPTER 5

1. N. Nahi and T. Assefi, Bayesian Recursive Image Estimation, *IEEE Trans. Comput.*, Vol. C-12, No. 7, July 1972.

2. T. Assefi, *Two-Dimensional Signal Processing with Application to Image Restoration*, Technical Report 32-1596, Jet Propulsion Laboratory, Pasadena, CA, Sept. 1, 1974.

3. T. Assefi, Modeling of Two-Dimensional Signals with Application to Image Restoration, *Proc. Third Intern. Joint Conf. Pattern Recognition*, IEEE Computer Society, pp. 696–700, Nov. 8-11.

4. C. Franco, Recursive Image Estimation, Ph.D. thesis, University of Southern California, 1973.

5. B. Anderson and J. Moore, Spectral Factorization of Time Varying Covariance Functions, *IEEE Trans. Inform. Theory*, Vol. IT-15, pp. 550-557, Sept. 1969.

6. T. Kailath, Covariance Factorization — An Explication via Examples, Conference Record of Second Asilomar Conference on Circuits and Systems, October 30-November 1, 1968.

7. S. Powell and L. Silverman, Modeling of Two-Dimensional Random Fields with Application to Image Restoration, *IEEE Trans. Autom. Control*, Feb. 1974.

8. N. Weiner, *The Extrapolation, Interpolation, and Smoothing of Stationary Time Series with Engineering Applications*, Wiley, 1949.

9. L. Franks, A Model for Random Video Signals, *Bell Syst. Tech. J.*, Vol. 45, pp. 609–630, Apr. 1966.

10. B. Anderson and J. Moore, The Kalman-Bucy Filter as a True Time Varying Weiner Filter, *IEEE Trans. Syst., Man, Cybernetics*, Vol. SMC-1, pp. 119–128, Apr. 1971.

11. L. Franks, *Signal Theory*, Prentice-Hall, 1969.

12. L. Rhodes, A Tutorial Introduction to Estimation and Filtering, *IEEE Trans. Control Theory*, Vol. AC-16, pp. 688–706, Dec. 1971.

13. N. Nahi, *Estimation Theory and Its Applications*, Wiley, 1969.

14. E. Wong, *Stochastic Processes in Information and Dynamical Systems*, McGraw-Hill, 1971.

15. H. Cramer and H. Leadbetter, *Stationary and Related Stochastic Processes*, Wiley, 1967.

ADDITIONAL REFERENCES FOR FURTHER READING

16. M. Murphy and L. Silverman, Image Model Representation and Line-by-Line Recursive Restoration, *IEEE Trans. Autom. Control*, Vol. AC-23, Oct. 1978.

17. O. Aboutalib and L. Silverman, Restoration of Motion Degraded Imaged Degraded by Blurs, *IEEE Trans. Circuits Syst.*, Vol. CAS-22, pp. 278-286, March 1975.

18. A. Habibi, Two-Dimensional Bayesian Estimate of Images, *Proc. IEEE*, Vol. 60, pp. 678-883, July 1972.

19. E. Wong, Detection and Filtering for Two-Dimensional Random Fields, *Proc. Conf. Decision and Control*, pp. 591-595, Dec. 1976.

20. A. Jain and E. Angel, Image Restoration, Modeling and Reduction of Dimensionality, *IEEE Trans. Computers*, Vol. C-23, pp. 470-476, May 1974.

21. V. Heine, Models for Two Dimensional Stationary Stochastic Processes, *Biometrika*, Vol. 42, pp. 170-178, 1955.

22. E. Wong, Recursive Causal Linear Filtering for Two Dimensional Random Fields, *IEEE Trans. Information Theory*, Vol. IT-24, pp. 50-59, Jan. 1978.

23. A. Jain, Some New Techniques in Image Processing, *Proc. Symp. on Image Science Mathematics*, Nov. 1976.

24. J. Woods, Markov Image Modeling, *IEEE Trans. Automatic Control*, Vol. AC-23, pp. 846-850, Oct. 1978.

25. E. Angel and A. Jain, Frame-to-Frame Restoration of Diffusion Images, *IEEE Trans. Automatic Control*, Vol. AC-23, pp. 850-855, Oct. 1978.

INDEX

Anderson, 285, 286
Assefi, 286, 287
Autocovariance (see Covariance)
Average Power, 83

Band-pass process, 110-116
Band-limited Processes, 100, 109
Bayes' theorem, 3
Bilateral Laplace transform, 86-87 App. C
Bochner's theorem, 93
Bucy, 121, 158, 285

Calculus of variation, 232
Causal systems, 77, 145
Characteristic functions, 29
Combination of estimates, 183-184
Complex random variables, 17
Conditional density functions, 9
Conditional probability, 2
Convergence, 18
Convolution, 78
Correlation coefficient, 17, 47
Correlation function, 46
Covariance, 17-18, 28
Cross-correlation, 47
Cross-covariance, 47

Delta function, 77 App. A
Deutsch, 285
Dirichlet condition, 80, App. C
Distribution function, 5

Empty Set, 2
Ergodicity, 56, 63

Error function, 19
Estimation theory, 120
Events, 1
Expected values, 16
Experiments, 1, 2

Franco, 287
Franks, 287
Filtering, 162
First-order statistics, 40
Fourier series, 80, App. C and D
Fourier transform, 81, App. C
Fourier transform pair, 82, 83, App. C
Functions of random variables, 10-15

Gaussian functions, 7
Gnedenko, 284

Hilbert transforms, 112
Holder inequality, 17

Image statistics, 213
Impulse function, 77, App. A
Impulse response, 77
Independent events, 4
Independent increments, 64
Instaneous power, 83
Integrals, see Stochastic integrals
Interpolation function, 103
Ito, 64

Jocabian, 15
Joint density, 8
Joint distribution, 8

Kailath, 178, 286
Kalman, R. E., 121, 158-192, 283, 284
Kalman-Bucy filtering 157-191

Laplace, see bilateral Laplace transforms
Lathi, 284
Leadbetter, 283
Levy process, 68
Liebelt, 285
Linear estimate, 126
Linear-mean-square estimation, 126, 134
Linear systems, 75
Linear transformation, App. B
Low-pass signals, 108

Marginal density function, 8-9
Marginal distribution function, 8-9
Markov processes, 157
Matched Filtering, 154
Mean, 16
Mean-square, continuity, 53
Mean-square-estimation, 123-126
Meyer, 283
Moore, 286
Mutual events, 2

Nahi, 284, 287
n-dimensional space, App. B
Noisy observations, 120
Nonlinear systems, 122
Norm, 129, App. B
Normal random variables, 7
nth order density, 14
nth order distribution, 14

Operator, 74
  linear, App. B
Optimum filter, 136
Orthogonal process, 53, 99
Orthogonality principle, 128
Outcomes, 1, 8

Papoulis, 283, 284
Partial-randomization, 228
Partitioned image, 228
Periodic processes, 104
Poisson process, 64-65
Positive definitie, App. B
Powell, 287
Power spectrum, 74, 83, 86
Power spectral density, 84
Prediction, 150, 179

Probability density function, 4-10, 40
Probability distribution function, 4-10, 40

Quadratic mean, 18
Quadrature component, 116

Random variables, 4
  complex, 17
  discrete, 5
  periodic, 104
Random vectors, 29
Rayleigh density function, 7
Recursive filtering (see Kalman-Bucy filtering)
Recursive image estimation, 202
Realizable systems, 77
Rhodes, 285
Risk function, 222

Sample functions, 36
Sample space, 1
Sampling theorem, 100
Scanner, 202
Schwarz, 284
Schwarz inequality, 17
Second-order statistics, 45
Silverman, 287
Shanon, 100
Signal-to-noise ratio, 156, 225
Smoothing, 150, 184
Spectral factorization, 204
State-variables, 123, App. E
Standard deviation, 16
Stationary correlation function, 48-49
  wide-sense, 49
Stieltjes integrals, 16
Stochastic continuity, 53
Stochastic differentiation, 53-55
Stochastic integration, 56-60
Stochastic processes, 36-39
Systems
  dynamics, 77
  instaneous, 77
  lumped, 77
  memory, 77
Systems and modeling, 121-122

Time averages, 63
Time-invariant systems, 78
Transition matrix, 206, App. E
Transformation linear, App. B
Two-dimensional signals, 202

Unbiased estimates
   conditional, 128
Unconditional, 172
Uncorrelated processes, 46
Uncorrelated random variables, 36
Uniform density functions, 7

Variance, 16
Vector spaces, 129, App. B

White noise, 68
Wide-sense stationarity, 49
Wiener, 121
Wiener process, 64, 67
Wiener-Hopf equation, 138-139, 145
Wiener-Kolmogorov theory, 137, 144
Wong, 283, 284